An Introduction To

DIFFERENTIAL MANIFOLDS

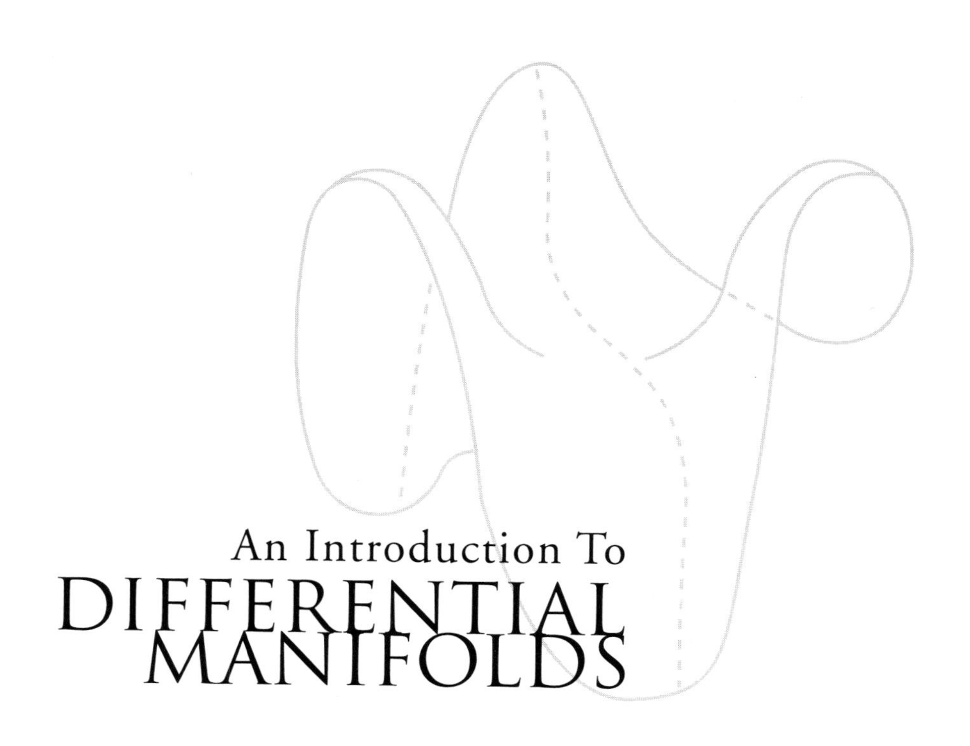

An Introduction To DIFFERENTIAL MANIFOLDS

Dennis Barden & Charles Thomas
University of Cambridge, UK

ICP Imperial College Press

Published by

Imperial College Press
57 Shelton Street
Covent Garden
London WC2H 9HE

Distributed by

World Scientific Publishing Co. Pte. Ltd.
5 Toh Tuck Link, Singapore 596224
USA office: 27 Warren Street, Suite 401-402, Hackensack, NJ 07601
UK office: 57 Shelton Street, Covent Garden, London WC2H 9HE

British Library Cataloguing-in-Publication Data
A catalogue record for this book is available from the British Library.

First published 2003
Reprinted 2005

AN INTRODUCTION TO DIFFERENTIAL MANIFOLDS

ISBN 1-86094-354-3
ISBN 1-86094-355-1 (pbk)

Printed in Singapore by World Scientific Printers (S) Pte Ltd

PREFACE

There are two ways to think of a differential manifold M^m. In the first, which is more abstract, M^m is the union of patches (or coordinate neighbourhoods) each identifiable with an open subset of $\mathbb{R}^m$, the patches being glued together by differentiable maps. In the second we regard M^m as being a subset of some higher dimensional space $\mathbb{R}^n$, defined for example by the vanishing of a family of differentiable functions $f : \mathbb{R}^n \to \mathbb{R}$. As we shall see, these two approaches are actually equivalent. The second, historically earlier, approach was used by Gauss in his study of the curvature properties of surfaces ($m = 2$), immersed in $\mathbb{R}^3$, while the first was tentatively formulated for arbitrary dimensions by Riemann in his famous inaugural lecture in June 1854. Much of this lecture was devoted to the metric problems considered by Gauss, and these continued to dominate the subject for some time. It was only as a result of the development of a topological as opposed to a metric framework in the first third of the twentieth century that *differential topology*, that is the study of manifolds independently of any particular distance function, really took off. The assumption of differentiability (or smoothness) leads to theorems which are either false or much harder to prove for those manifolds for which we ask only that the local patches overlap continuously (C°). As often happens in mathematics generalisation leads to simplification; certains properties turn out to be independent of any particular metric structure. For example the structure of N^n near an embedded submanifold M^m depends only on M^m and a certain 'normal bundle'. This can be shown 'geometrically', that is by exploiting the properties of some 'Riemannian metric' imposed on the ambient manifold. But on closer inspection all we really need to use is the theory of second order ordinary differential equations satisfying a suitable condition (sprays).

Riemannian geometry is only one way of restricting the general theory of differential manifolds. This has been elegantly expressed by Michael Atiyah † in the following way:

A real vector space $\mathbb{R}^{2n}$ can be given any one of the following extra structures [complex, Euclidean, symplectic] with the associated automorphism group [$GL(n, \mathbb{C})$, $O(2n)$, $Sp(n)$] $\cdots$ Any two of these

† M.F. Atiyah, *The Moment Map in Symplectic Geometry*, Durham Symposium on Global Riemannian Geometry, Ellis Horwood Ltd. (1984), 43–51.

structures which are compatible, in an obvious sense, determine the third and give altogether a Hermitian structure. The corresponding group $U(n)$ is $\cdots$ the maximal compact subgroup both of $GL(n, \mathbb{C})$ and $Sp(n)$. $\cdots$ each of these linear structures leads to its appropriate non-linear geometry. Thus we get (1) complex analytic geometry (2) Riemannian geometry and (3) symplectic geometry. $\cdots$ The three geometries intersect in Kähler geometry, the importance of which lies in the fact that all projective algebraic varieties are Kähler.

We have quoted this at some length partly to emphasise the importance of the 'infinitesimal model' in geometry. Near any point of M^{2n} the manifold is approximated by its tangent space, and the union of all these forms the tangent bundle, the reducibility of whose structure group is the first clue to the existence of a finer geometric structure. And if Kähler geometry represents the intersection of three basic theories above, then differential topology represents their union, corresponding to the full linear automorphism group $GL(2n, \mathbb{R})$. To return to the original work of Gauss and Riemann, we can ask whether M^m embeds in some finite dimensional real linear space (the answer is 'yes' with embedding dimension $2m + 1$). Or we can ask whether M^m with some prescribed metric embeds in some $\mathbb{R}^n$ with the usual Euclidean metric, so generalising Gauss' original questions about surfaces immersed in dimension three. The answer is again 'yes', but the embedding dimension is much higher $(3n+11)n/2$. ‡ The reader of this book will meet techniques which are valuable in any more geometric setting, but which also contribute to the topological programme of classifying manifolds up to some wider equivalence relation.

In Chapter 1 the reader is introduced to differential manifolds and basic techniques. Important among these are variants of the inverse function theorem, and 'partitions of unity', a tool for passing from local to global results.

Chapter 2 is devoted to the 'infinitesimal theory', which by means of the important example of the tangent bundle TM introduces the reader to more general fibre bundles. These are explained in Chapter 3, and used to give examples of smooth manifolds in dimensions 3, 5 and 7. It is interesting to note that 'fibrations with finitely many singular fibres' are playing an important part in the contemporary reintroduction of geometry into topology. This holds particularly in dimension 3, where the fibrations which we describe may come close to exhausting the irreducible types. A major theme of the subject, and so of

‡ J. Nash, The Imbedding Problem for Riemannian Manifolds, Annals of Mathematics, **63** (1956) 20–63.

this book, is the extension of local constructions and procedures, originating in Euclidian space to global analogues on manifolds. Usually there is an obstruction to generalisation, although in Chapter 2 we were able to construct the derivatives of the differentiable maps that we defined in Chapter 1 without any further constraint. In Chapter 4 we carry out the slightly more ambitious globalisation of integration. For this we need to introduce differential forms on manifolds and also to ask that our manifolds be oriented. In Chapter 5, after extending the concept of the differential of a function to a derivation of the full algebra of forms, and also allowing our manifolds to have boundaries, we prove Stokes' Theorem. This, as well as generalising the fundamental theorem of the integral calculus and the special cases that the reader will have met in dimensions 2 and 3, leads naturally to Chapter 6 on de Rham cohomology. This is the core of the book and we devote a substantial amount of space to the calculation of the de Rham cohomology groups for some familiar spaces. Indeed after reading the sections on Morse functions and handle decompositions in Chapter 7, the reader should be able to prove directly that $H^k_{DR}(M)$ is isomorphic to $Hom_{\mathbb{R}}(H_k(M,\mathbb{R}),\mathbb{R})$, with H_k calculated from either simplexes or cells. We have not included this explicitly in the text, since this is a book on differential rather than algebraic topology. Other applications in Chapter 7 include the degree of a differentiable map, the index of a vector field, and the relation between this and the Euler number $\chi(M)$ (Hopf-Poincaré theorem). There is a short final chapter on Lie groups, which concludes with a discussion of their de Rham cohomology. Both this chapter and Chapter 3, as well as a few sections that are starred in the list of contents, may be omitted at first reading without risk of losing the thread of the text, though we hope they will inspire the reader to delve deeper into the subject.

Prerequisites are courses in linear algebra and analysis of one and several real variables. A first course on general topology would be useful but not essential. We have included a chapter setting out, with at least skeletal proofs, the theorems from differential analysis and algebra which we use at various points in the text. This is not meant to be exhaustive, but can be consulted at need, and is intended to make the reader's task easier.

Having read through the book and worked as many as possible of the exercises, the reader should be ready either for one of the geometries mentioned in the quotation from Atiyah, or for a more advanced course in differential topology, singularities or dynamical systems. So far as the first of these subjects go, besides the rival texts mentioned in our final literature survey, there are the published lecture notes of J. Milnor (1958) and the books of W. Browder (1972) and C.T.C. Wall (1999) on Surgery Theory. Although the former is easier to read, the latter, particularly in its updated edition, is more important as a pointer to future research. Both of us remember, as graduate students, attending an extended series of lectures by Wall, starting with the definition of a differential manifold and ending with the classification of free involutions on S^5. To the best of our knowledge Part III of these lectures, on immersions and embeddings, was never given, which means that the reader is forced back to the original papers. This part of the theory, which is represented classically by M.W. Hirsch's paper on immersions (1959), is relevant both for the classification of smooth manifolds, and in the construction of various 'geometric structures'. The reader is referred to either Gromov (1986) or Eliashberg and Mishachev (2002) for more on this subject, now known as the 'h-principle'.

Our book is based on lectures given by the authors over more than two decades as part of the senior year undergraduate course (Part 2 of the Mathematical Tripos) here in Cambridge. The schedules for this course, even its name (!) have varied over the years, but the core content has remained constant. We hope that in this more polished, written version, we have dealt with at least some of the problems which have baffled students over the years. Our thanks are due to these same students, especially to Tim Haire and Ian Short who proof read a near final version; to our colleague Huiling Le for reading and commenting on most versions of the text as it developed, as well as helping us to make the notation uniform and showing us how to get better figures with 'xfig'; and to Michèle Bailey for urgent help with the typing as our deadline approached. We are also grateful to the Press at Imperial College London for taking an interest in our project, and for helping us at all stages to bring it to fruition.

Cambridge, September 2002.

CONTENTS

(*) Sections 2.4, 2.7 and 7.7 as well as Chapters 3 and 8 may be omitted at a first reading

(†) Chapter 9 is intended for reference as the reader finds necessary.

CHAPTER 1

DIFFERENTIAL MANIFOLDS AND DIFFERENTIABLE MAPS

As for M. Jourdain: [†]

"Par ma foi! Il y a plus de quarante ans que je dis de la prose sans que j'en susse rien,"

most readers will soon discover that they have already worked, implicitly, on manifolds. For example, consider the problem of finding the maximum of the function $g : \mathbb{R}^m \to \mathbb{R}$ subject to the constraints $f_i(x) = c_i$, $i = 1, \cdots, k$ where $k < m$, and the functions f_i are also real-valued functions on $\mathbb{R}^m$. Then, under certain hypotheses that we shall make explicit later, the set $\{x \mid f_i(x) = c_i, i = 1, \cdots, k\}$ is a differential manifold M and we are in fact looking for the maximum of the function $g\big|_M$ on this manifold.

1.1. Differential Manifolds: Definitions and Examples.

Just as one selects the hypotheses that characterise transformation groups to define abstract groups and then shows that every such abstract group is in fact a transformation group, to define our differential manifolds we shall select the properties that allow one to carry out analysis in the above mentioned and related contexts and then show that every such abstract differential manifold does in fact arise as a subset of a Euclidean space.

First we define what it means for a set to be a manifold and then what it means for it to be differential.

Definition 1.1.1. A *topological manifold of dimension m* is a Hausdorff, second countable topological space M^m that is locally homeomorphic with $\mathbb{R}^m$.

Here we recall that

[†] J-B Molière, *Le Bourgeois Gentilhomme* (1671)

(*i*) *Hausdorff* means that any two distinct points lie in disjoint open sets;

(*ii*) *second countable* means that there is a countable family of open subsets such that each open subset is the union of a subfamily;

(*iii*) *locally homeomorphic with* $\mathbb{R}^m$ means that every point has an open neighbourhood homeomorphic with an open subset of $\mathbb{R}^m$.

The requirements that a manifold be Hausdorff and second countable are to some extent for technical convenience. Local homeomorphism with $\mathbb{R}^m$ is the crucial property that characterises a manifold.

Examples 1.1.2. For $m = 1$: the circle $\mathbf{S}^1$, the real line $\mathbb{R}^1$.

For $m = 2$: the 2-sphere $\mathbf{S}^2$, the torus $\mathbf{S}^1 \times \mathbf{S}^1$, the Möbius band without its boundary circle, the Klein bottle, the Cartesian plane $\mathbb{R}^2$. Of these, the torus and Möbius band are illustrated in Figure 1.1: the former may be represented by the inner tube for a cycle tyre with its valve removed and the resulting hole patched over; the latter is obtained by taking a strip of paper and gluing its two ends together after a single twist or, for that matter, any odd number of twists.

For $m = 3$: the real 3-dimensional space $\mathbb{R}^3$, an open subset U of $\mathbb{R}^3$, the 3-sphere $\mathbf{S}^3$.

For arbitrary m: the real m-dimensional space $\mathbb{R}^m$. We shall prove below that the m-sphere $\mathbf{S}^m$ is a compact m-manifold, a submanifold of $\mathbb{R}^{m+1}$, for all values of the natural number m.

The examples above will all turn out to be differential, rather than just topological, manifolds. The torus and Möbius band are the basic building blocks for *all* surfaces or 2-manifolds. The situation in dimension three is already far more complicated and not yet fully understood. We shall be able to give a more exciting range of examples at the end of Chapter 3.

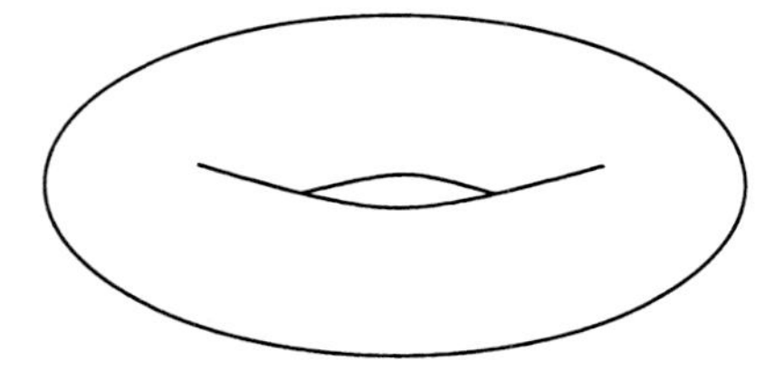
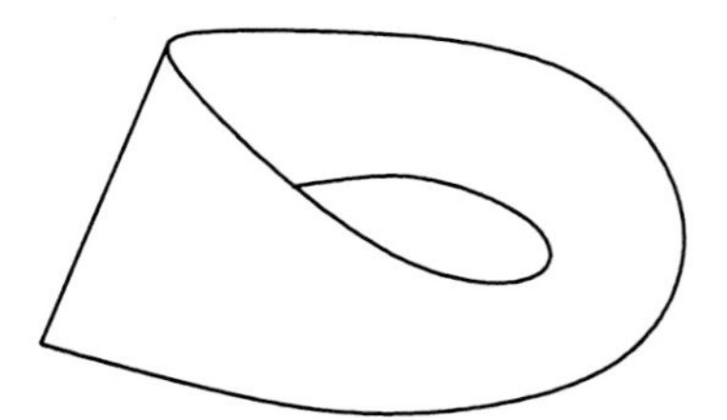

Figure 1.1.

The local homeomorphism of a manifold with $\mathbb{R}^m$ leads to charts for finding one's way around the manifold, just as one would on the earth's surface. Hence the following terminology.

Definition 1.1.3. A *chart* on a topological manifold M^m is a pair (ϕ, U) where U is an open subset of M and ϕ is a homeomorphism of U with an open subset $V = \phi(U)$ of $\mathbb{R}^m$. U is called a *coordinate neighbourhood* and V its *coordinate space.* If $0 \in V$, we say that the chart is *centred on* $p = \phi^{-1}(0) \in M$.

Each chart provides local coordinates; those of $\phi(q) \in \mathbb{R}^m$, for all q in U. Usually one chart will not cover the entire manifold so we need to be able to move between different coordinate systems. Returning to our geographic analogy, if we were given two maps of overlapping regions of the country this would be a simple matter: since it is conventional, other than formerly in China, always to put North at the top then, provided the maps are to the same scale, we only have to slide one of them so that the parts of the two maps that represent the common ground lie directly over one another. However, to correlate the representations under two manifold charts of the intersection of their domains, we need to allow general homeomorphisms, rather than just rescaling and translating. Given two such charts on a manifold, it is convenient for clarity to think of the images as lying in distinct copies of $\mathbb{R}^m$, distinct sheets of m-dimensional paper if you like, as in Figure 1.2.

Then

$$\theta_{12} = \phi_1 \circ \left(\phi_2\big|_{U_1 \cap U_2}\right)^{-1} : \phi_2(U_1 \cap U_2) \longrightarrow \phi_1(U_1 \cap U_2) \subseteq \mathbb{R}^m$$

and

$$\theta_{21} = \phi_2 \circ \left(\phi_1\big|_{U_1 \cap U_2}\right)^{-1} : \phi_1(U_1 \cap U_2) \longrightarrow \phi_2(U_1 \cap U_2) \subseteq \mathbb{R}^m$$

are inverse bijections. However $\left(\phi_2\big|_{U_1 \cap U_2}\right)^{-1} = \phi_2^{-1}\big|_{\phi_2(U_1 \cap U_2)}$ so it is continuous. Also $U_1 \cap U_2$ is open in U_1 and so $\phi_1(U_1 \cap U_2)$ is open in $\phi_1(U_1)$, since the homeomorphism ϕ_1 is an open mapping, and $\phi_1(U_1)$ is open in $\mathbb{R}^m$ so $\phi_1(U_1 \cap U_2)$ is also open in $\mathbb{R}^m$. Thus we have proved

Proposition 1.1.4. *The mappings θ_{12} and θ_{21} are inverse homeomorphisms between open subsets of $\mathbb{R}^m$. They are called coordinate transformations.* ■

The notation here will usually be abused, writing $\theta_{12} = \phi_1 \circ \phi_2^{-1}$, leaving the restriction of the domain implied. This is common practice in differential

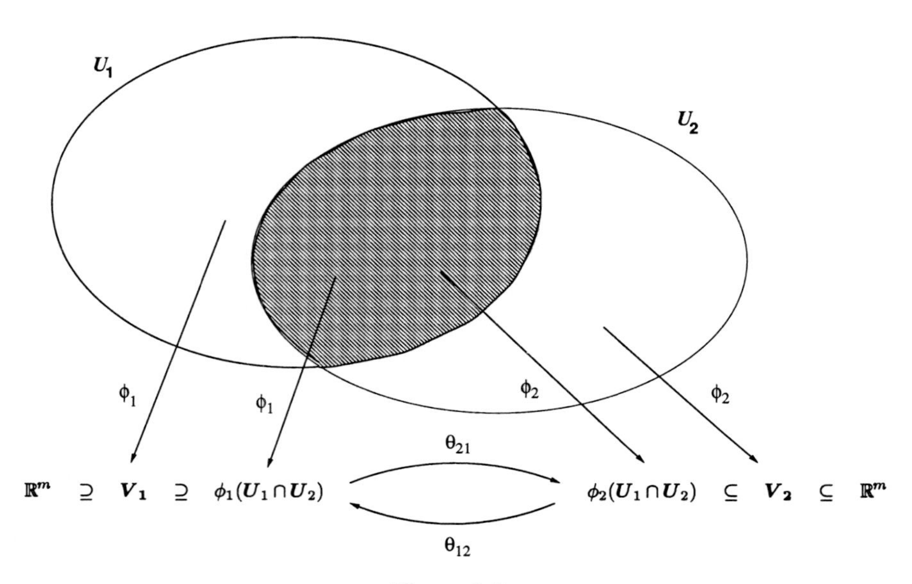

Figure 1.2.

geometry: suppression of gory details is often necessary to see the wood for the trees, particularly in computations. However this can make the subject opaque to the uninitiated. When in doubt the reader should always write in the domain of the map under consideration.

We are now ready for our basic definitions. The terminology continues to make geographic allusions.

Definition 1.1.5. A $\mathcal{C}^0$-*atlas for* M is a collection $\mathcal{A} = \{(\phi_\alpha, U_\alpha) \mid \alpha \in A\}$ of charts such that the coordinate neighbourhoods of members of $\mathcal{A}$ cover M: $\bigcup_{\alpha \in A} U_\alpha = M$.

Definition 1.1.6. A $\mathcal{C}^0$-atlas $\mathcal{A}$ is called a $\mathcal{C}^k$-*atlas* if all the coordinate transformations between members of $\mathcal{A}$ are of class $\mathcal{C}^k$, that is, they are k-times continuously differentiable. In the case $k = \infty$ the atlas, and the coordinate transformations, are called *smooth*.

> Throughout this book we shall assume that **all manifolds and related concepts are smooth**. Although we shall occasionally remind the reader of this, it should always be taken for granted unless we specify otherwise.

Temporary definition. A *smooth differential manifold* is a topological manifold M^m, together with a smooth atlas on M^m.

Examples 1.1.7. 1) Trivial: an open subset U of $\mathbb{R}^n$ with an atlas of just one chart id : $U \hookrightarrow \mathbb{R}^n$. In fact this is not quite as trivial as it looks since, for example, $GL(n, \mathbb{R})$ is an open subset of $\mathbb{R}^{n^2}$: the determinant being a continuous function so that $\det^{-1}(\mathbb{R} \setminus \{0\}) = GL(n, \mathbb{R})$ is open.

Note that the technical conditions, Hausdorff and second countable, are well-known in this case: any metric space is Hausdorff and any open subset of $\mathbb{R}^n$ is second countable. It is straightforward to show that any topological space which is a countable union of second countable open subspaces is itself second countable. In particular any (finite-dimensional) topological manifold with a countable atlas is second countable. These observations will enable us to check those technical requirements in most cases that we come across.

2) $\mathbf{S}^n = \{x = (x_0, \cdots, x_n) \in \mathbb{R}^{n+1} \mid \|x\| = 1\} \subseteq \mathbb{R}^{n+1}$ with the subspace (metric and) topology, and two charts:

$$\phi_N : U_N = \mathbf{S}^n \setminus \{N = (1, 0 \cdots, 0)\} \longrightarrow \mathbb{R}^n; \quad x \mapsto \frac{(x_1, \cdots, x_n)}{(1 - x_0)},$$

$$\phi_S : U_S = \mathbf{S}^n \setminus \{S = (-1, 0, \cdots, 0)\} \longrightarrow \mathbb{R}^n; \quad x \mapsto \frac{(x_1, \cdots, x_n)}{(1 + x_0)}.$$

To find the corresponding coordinate transformation we note that, for any point P on the sphere, other than N and S, the points P, $\phi_N(P)$ and $\phi_S(P)$ are coplanar with N, S and the centre O of the sphere, as illustrated in Figure 1.3. Then the similar triangles $OS\phi_S(P)$ and $O\phi_N(P)N$ show that $\|\phi_S(P)\| \, \|\phi_N(P)\| = 1$. So

$$\theta_{SN}(y) = y/\|y\|^2,$$

defined on the whole of $\mathbb{R}^n \setminus \{0\}$. This is $\mathcal{C}^\infty$ so we do have a smooth manifold.

3) In the special case $n = 2$ of the previous example we can identify $\mathbb{R}^2$ with $\mathbb{C}$ and call the coordinates z. Then $\theta_{SN}(z) = z/\{z\bar{z}\} = 1/\bar{z}$ and the new chart $\phi_2 = \mathrm{conj} \circ \phi_S : p \mapsto \overline{\phi_S(p)}$, defined on U_S, has $\theta_{2N}(z) = 1/z$. Thus we have

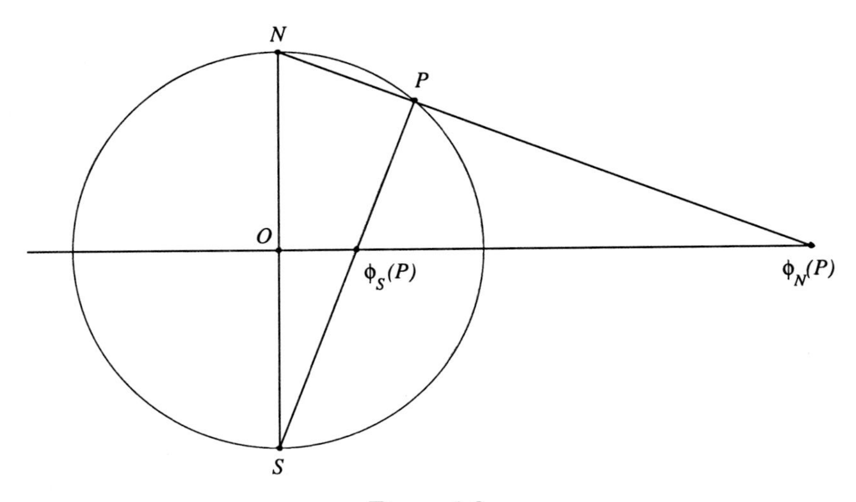

Figure 1.3.

two charts on the sphere having a complex analytic coordinate transformation between them. This makes it a complex analytic manifold, usually known as the Riemann sphere. Note that for the coordinate transformations to be complex analytic is much stronger than the real $\mathcal{C}^\infty$ condition of Definition 1.1.6.

But wait: we now have rather too many differential manifolds. We could have an atlas on $\mathbf{S}^n$ of two charts different from those chosen above, for example projecting from any other pair of antipodal points. Do we really wish to regard this as a different manifold? Well no, but to avoid it requires a logical detour.

Definition 1.1.8. We say two charts (ϕ_α, U_α) and (ϕ_β, U_β) are (smoothly) *compatible* if the coordinate transformation $\theta_{\alpha\beta} = \phi_\alpha \circ \phi_\beta^{-1}$ is smooth.

Thus the definition of an atlas may be rephrased as a family of compatible charts whose coordinate neighbourhoods form an open cover of M^m.

Definition 1.1.9. An atlas $\mathcal{A}$ is *maximal* if each chart (γ, W), which is compatible with every chart of $\mathcal{A}$, is already in $\mathcal{A}$.

Lemma 1.1.10. *Each (smooth) atlas $\mathcal{A}$ on a topological manifold M^m is contained in a unique maximal (smooth) atlas $\mathcal{M}$.*

Proof. Let

$$\mathcal{M} = \{(\psi, V) \mid (\psi, V) \text{ is a chart compatible with all charts of } \mathcal{A}\}.$$

Then:

(a) Firstly $\mathcal{M}$ *is* an atlas. For, if p is any point common to the domains of two charts (ϕ_1, U_1) and (ϕ_2, U_2) of $\mathcal{M}$ and (ϕ, U) is a chart about p from $\mathcal{A}$, then both the former charts are compatible with the last one. So $\phi_1 \circ \phi^{-1}$ is differentiable at $\phi(p)$ and $\phi \circ \phi_2^{-1}$ is differentiable at $\phi_2(p)$. Thus $\phi_1 \circ \phi_2^{-1}$ is differentiable at $\phi_2(p)$ also. Similarly $\phi_2 \circ \phi_1^{-1}$ is differentiable at $\phi_1(p)$. Since p is an arbitrary point common to the domains of the two charts (ϕ_1, U_1) and (ϕ_2, U_2), they are compatible.

(b) $\mathcal{M} \supseteq \mathcal{A}$, since $\mathcal{A}$ is an atlas.

(c) $\mathcal{M}$ is maximal. For if (ν, W) is a chart compatible with all charts of $\mathcal{M}$ then, in particular, it is compatible with all charts of $\mathcal{A}$ and so, by definition, in $\mathcal{M}$.

(d) If $\mathcal{N}$ is any atlas containing $\mathcal{A}$, then $\mathcal{N} \subseteq \mathcal{M}$. For each chart of $\mathcal{N}$ is compatible in particular with every chart of $\mathcal{A}$ and so is in $\mathcal{M}$.

(e) If $\mathcal{N}$ is a maximal atlas containing $\mathcal{A}$ then, as $\mathcal{N} \subseteq \mathcal{M}$, any $(\psi, V) \in \mathcal{M}$ is compatible with every chart of $\mathcal{N}$ and so, as $\mathcal{N}$ is maximal, is already in $\mathcal{N}$. Thus $\mathcal{M} \subseteq \mathcal{N}$.

Thus $\mathcal{M}$ is the unique maximal atlas containing $\mathcal{A}$. ■

Definition 1.1.11. A (smooth) *differential manifold M^m of dimension m* is a topological manifold of dimension m together with a *maximal* (smooth) atlas on it.

Remarks 1.1.12. 1) Here ends our logical detour. An alternative route is to define atlases to be equivalent if their union is an atlas and then define a differential structure on a manifold to be an equivalence class of atlases.

2) The maximal atlas on a differential manifold is also referred to as its *differential structure.*

3) We may now specify a differential structure on a topological manifold by *any* convenient small atlas, as we did for $\mathbf{S}^n$. In practice we shall always seek to minimise the number of charts in any atlas that we use to determine a differential structure.

Example 1.1.13. $\mathbb{R}\mathrm{P}^n$, real projective n-space, is the quotient of $\mathbb{R}^{n+1} \setminus \{0\}$ by the equivalence relation $x \sim y$ if there exists $\lambda \in \mathbb{R} \setminus \{0\}$ such that $y = \lambda x$.

We put the quotient topology on $\mathbb{R}\mathrm{P}^n$, which means that U is open in $\mathbb{R}\mathrm{P}^n$ if and only if $q^{-1}(U)$ is open in $\mathbb{R}^{n+1} \setminus \{0\}$ where q is the quotient map. Then, for two points x, y in $\mathbb{R}^{n+1} \setminus \{0\}$ not collinear with 0, we can find disjoint open sets X, Y in $\mathbb{R}^{n+1} \setminus \{0\}$, each a union of lines with the origin deleted, with $x \in X$ and $y \in Y$. Then $q^{-1}(q(X)) = X$ so $q(X)$ is open and similarly so is $q(Y)$. Thus $\mathbb{R}\mathrm{P}^n$ is Hausdorff.

We shall give $\mathbb{R}\mathrm{P}^n$ a differential structure determined by $n+1$ charts. Hence in particular it will be second countable. Let $V_i = \{x \in \mathbb{R}^{n+1} \setminus \{0\} \mid x_i \neq 0\}$. Then, $q^{-1}(q(V_i)) = V_i$ so $U_i = q(V_i)$ is open in $\mathbb{R}\mathrm{P}^n$ and we may define

$$\Phi_i : V_i \longrightarrow \mathbb{R}^n; \quad (x_0, \cdots, x_n) \mapsto \left(\frac{x_0}{x_i}, \cdots, \frac{x_{i-1}}{x_i}, \frac{x_{i+1}}{x_i}, \cdots, \frac{x_n}{x_i}\right)$$

which, being constant on equivalence classes, induces $\phi_i : U_i = q(V_i) \to \mathbb{R}^n$. Since the V_i cover $\mathbb{R}^{n+1} \setminus \{0\}$, their images, the U_i, cover $\mathbb{R}\mathrm{P}^n$.

Also, for U open in $\mathbb{R}^n$, $q^{-1}(\phi_i^{-1}(U)) = \Phi_i^{-1}(U)$ is open since Φ_i is continuous. Hence $\phi_i^{-1}(U)$ is open and ϕ_i is continuous.

Define

$$\Psi_i : \mathbb{R}^n \longrightarrow V_i; \quad (y_1, \cdots, y_n) \mapsto (y_1, \cdots, y_i, 1, y_{i+1}, \cdots, y_n).$$

Then $\phi_i \circ (q \circ \Psi_i) = \mathrm{id}_{\mathbb{R}^n}$ so $\phi_i^{-1} = q \circ \Psi_i$ which is continuous. Thus (ϕ_i, U_i) *is* a chart.

Then $\theta_{ji}(y_1, \cdots, y_n)$ is equal to

$$\begin{cases} \left(\frac{y_1}{y_{j+1}}, \cdots, \frac{y_j}{y_{j+1}}, \frac{y_{j+2}}{y_{j+1}}, \cdots, \frac{y_i}{y_{j+1}}, \frac{1}{y_{j+1}}, \frac{y_{i+1}}{y_{j+1}}, \cdots, \frac{y_n}{y_{j+1}}\right) & \text{if } j < i, \\ \left(\frac{y_1}{y_j}, \cdots, \frac{y_i}{y_j}, \frac{1}{y_j}, \frac{y_{i+1}}{y_j}, \cdots, \frac{y_{j-1}}{y_j}, \frac{y_{j+1}}{y_j}, \cdots, \frac{y_n}{y_j}\right) & \text{if } j > i, \end{cases}$$

which is smooth on

$$\phi_i(U_i \cap U_j) = \begin{cases} \{y \mid y_{j+1} \neq 0\} & \text{if } j < i, \\ \{y \mid y_j \neq 0\} & \text{if } j > i. \end{cases}$$

Thus $\mathbb{R}\mathrm{P}^n$ is a smooth manifold. In fact it is a real analytic manifold. The construction of a projective space can be carried out equally well for the complex numbers $\mathbb{C}$ and the quaternions $\mathbb{H}$. In the former case $\mathbb{C}\mathrm{P}^n$, of real dimension $2n$, is a complex analytic manifold, but the definition of a 'quaternionic manifold' presents difficulties that we shall not discuss here.

Example 1.1.14. *Product manifolds.* If M^m and N^n are smooth manifolds with atlases $\{(\phi_\alpha, U_\alpha) \mid \alpha \in A\}$ and $\{(\psi_\beta, V_\beta) \mid \beta \in B\}$ respectively then $\{(\phi_\alpha \times \psi_\beta, U_\alpha \times V_\beta) \mid \alpha \in A, \beta \in B\}$ is an atlas on $M \times N$, the topological product, which is therefore Hausdorff and second countable. The differential structure it determines depends only on those of M and N. It is straightforward, but rather tedious, to check this last statement.

Special case: $T^n = \mathbf{S}^1 \times \cdots \times \mathbf{S}^1$ with n factors is the n-fold torus.

1.2. Differentiable Mappings: Definitions and Examples.

As a first application of our putting differential structures on topological manifolds we may now define what it means for functions on, and maps between, manifolds to be differentiable and we may also define related concepts.

Definition 1.2.1. A function $f : M^m \to \mathbb{R}$ (or $f : M^m \to \mathbb{C}$, $f : M^m \to \mathbb{R}^n$) is *differentiable at p in M* if, for some chart (ϕ, U) at p, $f \circ \phi^{-1}$ is differentiable at $\phi(p)$.

Proposition 1.2.2. *This is well-defined.*

Proof. That is, it is independent of the choice of chart at p. For, if (ψ, V) is compatible with (ϕ, U), then

$$f \circ \psi^{-1} = f \circ \phi^{-1} \circ \phi \circ \psi^{-1} = f \circ \phi^{-1} \circ \theta_{\phi\psi}$$

on the appropriately restricted domain, which still however contains p. Since the coordinate transformation $\theta_{\phi\psi}$ is smooth, so too is $f \circ \psi^{-1}$. ■

Definition 1.2.3. The function $f : M^m \to \mathbb{R}$ is differentiable on the open set $U \subseteq M$ if it is differentiable at all points p in U.

Corollary 1.2.4. 1) *For a chart (ϕ, U) on M, f is differentiable on U if and only if $f \circ \phi^{-1}$ is differentiable on $\phi(U)$.*

2) *The function f is differentiable on M if and only if, for some atlas $\mathcal{A}$, $f \circ \phi^{-1}$ is differentiable on $\phi(U)$ for all $(\phi, U) \in \mathcal{A}$ which is if and only if, for all atlases $\mathcal{A}$, $f \circ \phi^{-1}$ is differentiable on $\phi(U)$ for all $(\phi, U) \in \mathcal{A}$.* ■

Definition 1.2.5. Similarly, a map $f : M^m \to N^n$ is *differentiable at $p \in M^m$* if it is continuous at p and, for suitable charts (ϕ, U) at p in M and (ψ, V) at $f(p)$ in N, $\psi \circ f \circ \phi^{-1} : \phi(U \cap f^{-1}(V)) \to \mathbb{R}^n$ is differentiable at $\phi(p)$. It is differentiable on an open subset U of M if it is continuous on U and if it is differentiable at all points of U.

The requirement of continuity is to ensure that $\psi \circ f \circ \phi^{-1}$ is defined on some neighbourhood of $\phi(p)$ so that it is meaningful to ask for differentiability. All we strictly need is that each point p at which we wish to check differentiability has a coordinate neighbourhood U such that $f(U)$ is contained in a coordinate domain on N. Hirsch (1976) adopts this as his definition, but most authors use the version we have given.

Similar corollaries apply for this extended definition as did for the special case $N = \mathbb{R}$, though now we must also allow for atlases on N.

Notation. We write $\mathcal{C}^\infty(M, N)$ for the set of all smooth mappings from M to N and abbreviate $\mathcal{C}^\infty(M, \mathbb{R})$ to $\mathcal{C}^\infty(M)$.

We note that the 'pointwise operations' in the image $\mathbb{R}$ turn $\mathcal{C}^\infty(M)$ into an algebra over $\mathbb{R}$: for $\lambda, \mu \in \mathbb{R}$ and $f, g \in \mathcal{C}^\infty(M)$

$$(\lambda f + \mu g)(x) = \lambda f(x) + \mu g(x) \quad \text{and} \quad (fg)(x) = f(x)\, g(x).$$

(See Chapter 9 for the definition of an algebra.)

Definition 1.2.6. A mapping $f : M^m \to N^n$ is called a *diffeomorphism* if it is a homeomorphism and both it and its inverse are differentiable.

Standard counterexample. A differentiable homeomorphism need not be a diffeomorphism:

$$f : \mathbb{R}^1 \longrightarrow \mathbb{R}^1; \quad x \mapsto x^3$$

is a smooth homeomorphism, but its inverse is not differentiable at 0.

Definition 1.2.7. The *rank at p in M* of the differentiable map $f : M^m \to N^n$ is the rank of the derivative $D(\psi \circ f \circ \phi^{-1})$ at $\phi(p)$, where ϕ is a chart on M at p and ψ a chart on N at $f(p)$.

This is defined since $\psi \circ f \circ \phi^{-1}$ is a differentiable map between open subsets of Euclidean spaces. It is well defined, as above, because the coordinate transformations on both manifolds are smooth diffeomorphisms and so their derivatives are isomorphisms.

Examples 1.2.8. 1) The product map:

$$GL(n, \mathbb{R}) \times GL(n, \mathbb{R}) \to GL(n, \mathbb{R}); (a, b) \mapsto ab$$

is differentiable.

2) The exponential map $\exp : \mathbb{R} \to \mathbf{S}^1; x \mapsto e^{ix}$ is differentiable.

3) Fix $g \in GL(n, \mathbb{R})$. Then the left translation

$$L_g : GL(n, \mathbb{R}) \to GL(n, \mathbb{R}); \; a \mapsto ga$$

is a diffeomorphism.

4) (*i*) The quotient map $q : \mathbb{R}^{n+1} \setminus \{0\} \to \mathbb{R}\mathrm{P}^n$ has rank n at all points.

(*ii*) The restriction: $\mathbf{S}^n \overset{q|_{\mathbf{S}^n}}{\longrightarrow} \mathbb{R}\mathrm{P}^n$ of the quotient map is a local diffeomorphism, but not a bijection, so it is not a global diffeomorphism.

1.3. Submanifolds and the Inverse Function Theorem.

We begin this section with the all-important Inverse Function Theorem and some of its associated results. Readers finding the outline proofs provided too brief for understanding should turn to the 'prerequisites' in Chapter 9.

Theorem 1.3.1. (Inverse Function Theorem.) *Let U be open in $\mathbb{R}^m$ and $f : U \to \mathbb{R}^n$ be differentiable and such that, for some $x \in U$, $Df(x)$ is an isomorphism. Then $m = n$ and f is a local diffeomorphism at x.*

Outline of proof. To be a local diffeomorphism at x means that there is an open subset V of U such that $x \in V$, $f(V)$ is open and $f : V \to f(V)$ is a diffeomorphism. Without loss of generality, we may assume that $x = 0$, $f(0) = 0$ and $Df(0) = \text{id}$, using obvious local diffeomorphisms to reduce the general case to this one.

Now use the Mean Value Theorem to show there exists r such that

$$g_y : x \mapsto x + y - f(x) = y + g_0(x)$$

is a contraction mapping on $\{x \mid \|x\| \leqslant r\}$ for all y such that $\| y \| \leqslant r/2$. By the Contraction Mapping Theorem, 9.2.2, there is a unique x in $\overline{B(0, r)}$, the closed ball of radius r, such that $g_y(x) = x$, which is equivalent to $f(x) = y$. Hence there is an inverse to f from the ball of radius $r/2$ into the ball of radius r. The continuity of f^{-1} follows from its definition, but differentiability requires the chain rule. ■

For the next result we introduce our standard notation for injections of factors into products and projections from products onto factors:

$$\iota_2 : \mathbb{R}^p \longrightarrow \mathbb{R}^m \times \mathbb{R}^p; \quad y \mapsto (0,\, y) \quad \text{and}$$
$$\pi_2 : \mathbb{R}^m \times \mathbb{R}^p \longrightarrow \mathbb{R}^p; \quad (x,\, y) \mapsto y,$$

respectively, with analogous notation for other factors.

Corollary 1.3.2. (Submersion Theorem.) *If $0 \in U \subseteq \mathbb{R}^n = \mathbb{R}^m \times \mathbb{R}^p$ and if $f : U \to \mathbb{R}^q$ is smooth with the second partial derivative $D_2 f(0)$ $(= Df(0) \circ \iota_2 : \mathbb{R}^p \to \mathbb{R}^q)$ a linear isomorphism, then $p = q$ and there is a local diffeomorphism ϕ at 0 in $\mathbb{R}^n$ such that $f \circ \phi = \pi_2$.*

Proof. Let $\psi(x, y) = (x, f(x, y))$. Then $D\psi(0, 0)$ has matrix

$$J\psi(0,0) = \begin{pmatrix} I_m & 0 \\ J_1 f(0,0) & J_2 f(0,0) \end{pmatrix},$$

where $J_i f(0,0)$ is the Jacobian matrix of $D_i f(0,0)$. Thus $J\psi(0,0)$ is an isomorphism and so, by Theorem 1.3.1, ψ is a local diffeomorphism at $(0,0)$. If ϕ is the local inverse from $W \subseteq \mathbb{R}^m \times \mathbb{R}^p \to V \subseteq U$, then every point w in W is $w = \psi(x,y) = (x, f(x,y))$ for some (x,y) in V. Then $f \circ \phi(w) = f \circ \phi \circ \psi(x,y) = f(x,y) = \pi_2(w)$. ■

Corollary 1.3.3. (Implicit Function Theorem.) *Continuing with the data as in the statement and proof of Corollary* 1.3.2, *if* $(0,0) \in W_1 \times W_2 \subseteq W$ *and* $f(0,0) = 0$ *then there is a unique function* $g : W_1 \to \mathbb{R}^p$ *such that* $f(x, g(x)) = 0$ *for all* x *in* W_1.

Proof. Uniqueness follows from $\psi(x, g(x)) = (x, f(x, g(x))) = (x, 0)$ and the fact that ψ bijects locally.

For existence we define $g(x) = \pi_2 \circ \phi(x, 0)$ and then check that

$$\begin{aligned} f(x, g(x)) &= f(x, \pi_2(\phi(x,0))) \\ &= f(\pi_1(\phi(x,0)), \pi_2(\phi(x,0))) \\ &= f \circ \phi(x,0) = \pi_2(x,0) = 0, \end{aligned}$$

where we used the fact that ϕ preserves the first coordinate, since its inverse ψ did. ■

Addendum. *The map* $\alpha : W_1 \to f^{-1}(0); x \mapsto (x, g(x))$ *is a local homeomorphism onto its image, an open neighbourhood of* $(0,0)$ *in* $f^{-1}(0)$. *Moreover, any two such are compatible charts on* $f^{-1}(0)$.

Proof. We see that

$$\alpha(x) = (x, \pi_2 \circ \phi(x,0)) = (\pi_1 \circ \phi(x,0), \pi_2 \circ \phi(x,0)) = \phi(x,0) = \phi \circ \iota_1(x).$$

So $f \circ \alpha(x) = f \circ \phi(x,0) = \pi_2(x,0) = 0$, and $\alpha(x) \in f^{-1}(0)$. Also

$$\begin{aligned} \alpha(W_1) &= \phi(W_1 \times \{0\}) = \psi^{-1}(W_1 \times \{0\}) \\ &= \psi^{-1}(W_1 \times \mathbb{R}^p) \cap \psi^{-1}(\mathbb{R}^m \times \{0\}) = \psi^{-1}(W_1 \times \mathbb{R}^p) \cap f^{-1}(0). \end{aligned}$$

So $\alpha(W_1)$ is open in $f^{-1}(0)$. Then

$$\pi_1 \circ \psi \circ \alpha = \pi_1 \circ \psi \circ \phi \circ \iota_1 = \pi_1 \circ \iota_1 = \text{id}.$$

Thus α is injective and so bijects onto its image $\alpha(W_1)$ and both $\alpha = \phi \circ \iota_1$ and its inverse $\pi_1 \circ \psi$ are continuous.

Moreover, for two such charts, $\alpha_2^{-1} \circ \alpha_1 = \pi_1 \circ \psi_2 \circ \phi_1 \circ \iota_1$ is a composite of smooth functions and so smooth. Thus the two charts are compatible. ■

The Implicit Function Theorem shows that, if $f : \mathbb{R}^m \times \mathbb{R}^p \to \mathbb{R}^p$ is such that f has maximal rank p at each point of $f^{-1}(0)$, then $f^{-1}(0)$ is a smooth manifold. Actually it shows more: for x in $f^{-1}(0)$ there is a chart $\psi : V \to W \subseteq \mathbb{R}^{m+p}$ such that $\psi|_{V \cap f^{-1}(0)}$ is a chart on $f^{-1}(0)$ mapping to $W \cap (\mathbb{R}^m \times \{0\})$. We can generalise this situation, replacing $\mathbb{R}^m$ by an m-manifold.

Definition 1.3.4. The subset $M \subseteq N^{m+p}$ is a *smooth m-submanifold* or *codimension-p-submanifold* if for all x in M there is a chart $\phi : U \to \mathbb{R}^{m+p}$ for N about x such that $\phi(U \cap M) = \phi(U) \cap (\mathbb{R}^m \times \{0\})$. We refer to ϕ as a chart for N *adapted to* M.

Corollary 1.3.5. *An m-submanifold is an m-dimensional manifold.*

Proof. Argue as in the Addendum to Corollary 1.3.3 above. ■

Definition 1.3.6. For a map $f : M^m \to N^n$, a point x in M at which f has rank less than n is called a *critical point* and its image a *critical value* of f. Other points y in N, that is, such that f has rank n at all points x in $f^{-1}(y)$, are called *regular values* of f. If f is surjective and has rank n everywhere it is called a *submersion.*

Corollary 1.3.7. *If y is a regular value of $f : M^m \to N^n$, then $f^{-1}(y)$ is an n-codimensional, that is, an $(m-n)$-dimensional, submanifold of M.*

Proof. The definition of a submanifold is local, so it suffices to work on a coordinate neighbourhood or, equivalently, its coordinate space. But then we are mapping between subsets of $\mathbb{R}^m$ and $\mathbb{R}^n$ and we can use the Submersion and Implicit Function Theorems at each point of $f^{-1}(y)$ in that neighbourhood, or its image in the coordinate space. Note that, if $Df(x)$ has rank n and so is a surjection, we may precede it by a linear isomorphism of $\mathbb{R}^m$ to achieve the hypothesis of Corollary 1.3.2. ■

Remarks 1.3.8. Note that $f : M^m \to N^n$ can only have regular values if $m \geqslant n$. But then it is a consequence of Sard's Theorem, 9.5.4, that regular values abound for most maps and produce a plethora of submanifolds. See, for example, the section on Morse functions in Chapter 7. In particular we have submanifolds in $\mathbb{R}^n$. Conversely we shall see in the next section that, up to diffeomorphism, all manifolds so arise.

Examples 1.3.9. 1) The map norm: $\mathbb{R}^{n+1} \to \mathbb{R}$ is smooth on $\mathbb{R}^{n+1} \setminus \{0\}$ and all values in $\mathbb{R}^+$ are regular. So for all $r > 0$ the set $\text{norm}^{-1}(r)$ is a submanifold of $\mathbb{R}^{n+1}$ of dimension n, namely the sphere of radius r.

2) The map det: $GL(n, \mathbb{R}) \to \mathbb{R}$ is differentiable and 1 is a regular value. So $SL(n, \mathbb{R}) = \det^{-1}(1)$ is a smooth submanifold of $GL(n, \mathbb{R})$ (and of $\mathbb{R}^{n^2}$).

3) The subgroup $O(n)$ of $GL(n, \mathbb{R})$ is a $\mathcal{C}^\infty$-manifold of dimension equal to $n(n-1)/2$. We leave the proof of this as an exercise at the end of this chapter. It is further discussed in Chapter 8.

Theorem 1.3.10. (Local Embedding Theorem.) *If U is an open subset in $\mathbb{R}^m$ and $f : U \to \mathbb{R}^m \times \mathbb{R}^p$ is smooth such that $0 \in U$, $f(0) = 0$ and $Df_1(0) \equiv D(\pi_1 \circ f)(0)$ is an isomorphism, then there exists a local diffeomorphism at 0, $\theta : V \to \mathbb{R}^m \times \mathbb{R}^p$, such that $\theta \circ f = \iota_1$ and $\theta(V \cap f(U)) = \theta(V) \cap (\mathbb{R}^m \times \{0\})$.*

Proof. This is similar to the Local Submersion Theorem. See Chapter 9 for the details. ∎

Definition 1.3.11. A smooth map $f : M^m \to N^n$ which has rank m everywhere is called an *immersion*. A diffeomorphism of M^m onto an m-dimensional submanifold of N^n is called an *embedding* of M in N.

Counterexample. An embedding is an injective immersion but an injective immersion need not be an embedding: $f : \mathbb{R}^1 \to \mathbf{S}^1 \times \mathbf{S}^1$; $t \mapsto (e^{it}, e^{\pi it})$ has a dense image which is not a submanifold. We shall return to this example in Chapter 8.

Lemma 1.3.12. *An immersion which is a homeomorphism onto its image is an embedding.*

Proof. Let $f : M^m \to N^{m+p}$ be such an immersion. Then we have to show first that $f(M)$ is a submanifold of N. Let ϕ be a chart at p in M and ψ at $f(p)$ in N. Then $\psi \circ f \circ \phi^{-1}$ has rank m at $\phi(p)$ and, by Theorem 1.3.10, there is a diffeomorphism θ of a neighbourhood V of $\psi(f(p))$ in $\mathbb{R}^{m+p}$ such that, on $W = \phi(f^{-1}(\psi^{-1}(V)))$, the composite $\theta \circ \psi \circ f \circ \phi^{-1}$ is the inclusion of W as $W \times \{0\}$ in $\mathbb{R}^m \times \mathbb{R}^p$ and $\theta(V \cap (\psi \circ f \circ \phi^{-1}(W))) = \theta(V) \cap (\mathbb{R}^m \times \{0\})$. Now $\phi^{-1}(W)$ is open in M and, since f is a homeomorphism onto $f(M)$, $f(\phi^{-1}(W))$ is open in $f(M)$. By definition it is also contained in $\psi^{-1}(V)$. So there is an open set U in N, which we may assume to be contained in $\psi^{-1}(V)$, such that $f(\phi^{-1}(W)) = U \cap f(M)$. Then $(\theta \circ \psi, U)$ is the required chart such that

$\theta \circ \psi(U \cap f(M)) = (\theta \circ \psi(U)) \cap (\mathbb{R}^m \times \{0\})$. For

$$\begin{aligned}\theta \circ \psi(U \cap f(M)) &\subseteq (\theta \circ \psi(U)) \cap (\theta \circ \psi \circ f \circ \phi^{-1}(W)) \\ &\subseteq (\theta \circ \psi(U)) \cap (\mathbb{R}^m \times \{0\})\end{aligned}$$

and, conversely,

$$\begin{aligned}\theta(\psi(U)) \cap (\mathbb{R}^m \times \{0\}) &\subseteq \theta(V) \cap (\mathbb{R}^m \times \{0\}) \\ &= \theta(V \cap (\psi \circ f \circ \phi^{-1}(W))) = \theta(V \cap \psi(U \cap f(M))) \\ &\subseteq \theta \circ \psi(U \cap f(M)).\end{aligned}$$

Recalling that the corresponding chart on the submanifold $f(M)$ is the restriction of the chart $(\theta \circ \psi, U)$ we see that, with respect to this and the chart ϕ on M, f is given by the identity mapping between the corresponding coordinate spaces. So it is certainly a diffeomorphism of M onto the submanifold $f(M)$. ■

The crucial step in the proof above is the identification of the set U whose intersection with $f(M)$ is the set $f(\phi^{-1}(W))$ that we already know about: there are not, as there were in the previous counterexample, further slices of $f(M) \cap U$ getting arbitrarily close to $f(\phi^{-1}(W))$.

An important special case of the lemma arises when the domain manifold M^m is compact. Then, an injective continuous map is open and, hence, a homeomorphism onto its image. (See Chapter 9.) So an injective immersion of a compact manifold is always an embedding.

1.4. Bump Functions and the Embedding Theorem.

Definition. 1.4.1. A *Bump function* on $\mathbb{R}^n$ is a smooth function which takes the value 1 on some neighbourhood of the origin and the value 0 outside some larger neighbourhood.

The existence of bump functions distinguishes smooth manifolds from analytic ones: the uniqueness theorem for analytic functions means that analytic bump functions cannot exist. However, when they do exist, bump functions are a major tool, making possible globally on an entire manifold many constructions that are possible locally on its coordinate neighbourhoods. Thus the family $\mathcal{C}^\infty(M)$ of smooth real-valued functions on M is very large. Our first example of their use will be to obtain a weak version of the Whitney Embedding Theorem, showing that all manifolds can arise, as in the previous section, as submanifolds of Euclidean space.

Lemma 1.4.2. *For all $r > 0$ there is a smooth function $\Psi_r : \mathbb{R}^n \to [0,1] \subseteq \mathbb{R}$ such that*

(*i*) $\Psi_r(x) = 1 \iff \|x\| \leqslant r$,

(*ii*) $\Psi_r(x) = 0 \iff \|x\| \geqslant 2r$.

Proof. First, define

$$\lambda : \mathbb{R} \longrightarrow \mathbb{R}; \quad t \mapsto \begin{cases} 0 & \text{if } t \leqslant 0, \\ \exp(-1/t^2) & \text{if } t > 0, \end{cases}$$

and check that it is smooth. Then, for $r > 0$, define

$$\phi_r(t) = \frac{\lambda(t)}{\lambda(t) + \lambda(r-t)} \in [0,1].$$

This is well-defined since $\lambda(t) + \lambda(r-t) > 0$ for all t and $\phi_r(t) = 0$ if and only if $t \leqslant 0$, while $\phi_r(t) = 1$ if and only if $\lambda(r-t) = 0$, that is, if and only if $t \geqslant r$. Next, define

$$\psi_r : \mathbb{R} \longrightarrow \mathbb{R}; \quad t \mapsto 1 - \phi_r(|t| - r).$$

Then ψ_r is differentiable since it is constant near $t = 0$ and otherwise a composite of smooth functions. Finally, the function

$$\Psi_r : \mathbb{R}^m \longrightarrow \mathbb{R}; \quad x \mapsto \psi_r(\|x\|)$$

has the properties that we require. The functions λ, ϕ_r and ψ_r are illustrated in Figure 1.4. ■

Using bump functions, we may establish the following special case of Whitney's Embedding Theorem.

Theorem 1.4.3. *Let M^m be a compact manifold. Then there is an embedding of M^m in $\mathbb{R}^n$ for some n.*

Proof. At each point p in M take a chart $\phi_p : U_p \to \mathbb{R}^m$, centred on p, and choose r_p such that $\phi_p(U_p) \supseteq B(0, 3r_p)$. Let $V_p = {\phi_p}^{-1}(B(0, r_p))$, $W_p = {\phi_p}^{-1}(\overline{B(0, 2r_p)})$ and define

$$f_p(x) = \begin{cases} \Psi_{r_p} \circ \phi_p(x) & \text{for } x \in U_p, \\ 0 & \text{for } x \in M \setminus W_p. \end{cases}$$

Then f_p is well-defined and smooth on each of these overlapping open sets. Thus f_p is differentiable on M: it is called a *bump function on M* and is supported on W_p.

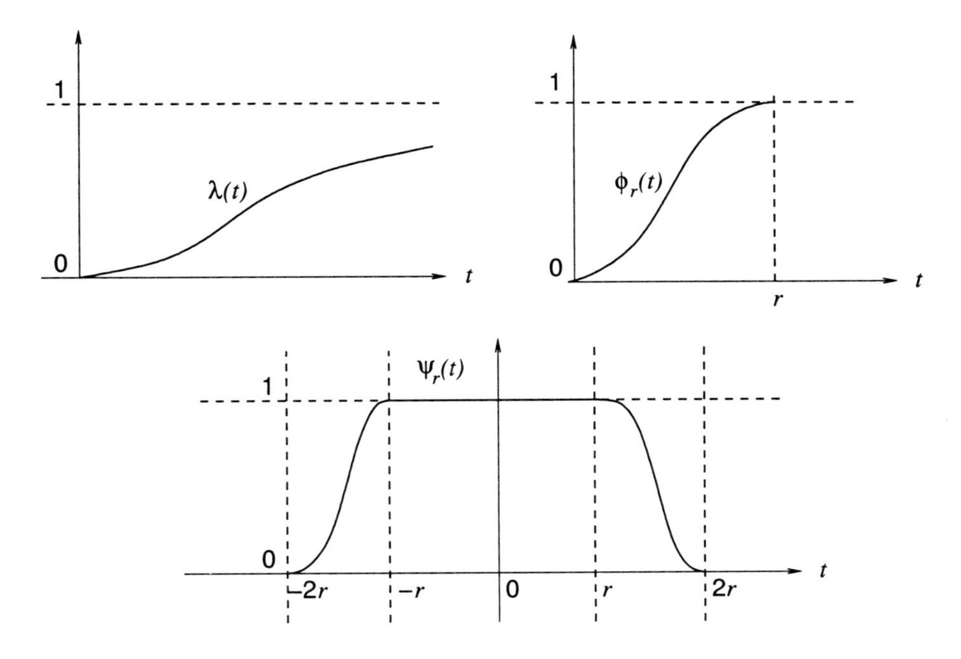

Figure 1.4.

Similarly, the mapping ψ_p from M into $\mathbb{R}^m$ defined by

$$\psi_p(x) = \begin{cases} f_p(x)\,\phi_p(x) & \text{for } x \in U_p, \\ 0 & \text{for } x \in M \setminus W_p \end{cases}$$

is smooth and suppported on W_p.

Since M is compact we may choose a covering of M by V_{p_i}, $i = 1, \cdots, k$. Then the mapping

$$f : M \longrightarrow \mathbb{R}^{km+k}; \quad x \mapsto (\psi_{p_1}(x), \cdots, \psi_{p_k}(x), f_{p_1}(x), \cdots, f_{p_k}(x))$$

is differentiable. It is also injective, since $x \neq y$ implies

either (i) there exists i such that $x, y \in \overline{V}_{p_i}$ and then

$$\psi_{p_i}(x) = \phi_{p_i}(x) \neq \phi_{p_i}(y) = \psi_{p_i}(y);$$

or else (ii) there exists i such that $x \in \overline{V}_{p_i}$ and $y \notin \overline{V}_{p_i}$ and then

$$f_{p_i}(x) = 1 \neq f_{p_i}(y).$$

Thus, in each case, the images $f(x)$ and $f(y)$ in $\mathbb{R}^{km+k}$, of the distinct points $x, y \in M^m$, are also distinct.

We also note that, if $x \in V_i$ and $\pi_i : \mathbb{R}^{km+k} \to \mathbb{R}^m$ is the projection onto the ith factor $\mathbb{R}^m$, then $\pi_i \circ f\big|_{V_i}$ is the homeomorphism ϕ_i which is the chart on V_i. So trivially $\pi_i \circ f$ is a diffeomorphism on V_i and, since M has dimension m, f must have rank m at x. Thus f is an injective immersion of a compact manifold and so an embedding. ■

For the actual theorem of Whitney, M does not need to be compact and the dimension n can be as low as $2m + 1$: we shall return to this topic at the end of the next chapter.

This result brings us full circle. Historically it was the Submersion, or Implicit Function, Theorem that led to the abstract definition of a manifold with which we opened this chapter. Disentangling the manifold from any particular embedding in Euclidean space has many advantages. However Whitney now tells us that it does not increase our stock of manifolds: up to diffeomorphism, every manifold is a submanifold of a Euclidean space. He also tells us that, when it simplifies matters, we may take our manifold to be so embedded. In most such cases there is no need for that embedding to be in the lowest possible dimension. The following is a typical example.

Corollary 1.4.4. *The manifold M admits a distance function d compatible with its topology.*

Proof. We simply take the Euclidean distance function restricted to M when it is embedded in a Euclidean space. ■

1.5. Partitions of Unity.

Although bump functions themselves were adequate for our crude version of the Embedding Theorem, they are more often applied in the form of partitions of unity defined as follows.

Definition 1.5.1. A family of functions $\{\sigma_\alpha \mid \alpha \in A\}$ on a manifold M is called a *partition of unity on M* if, for all $x \in M$, the sum $\sum\limits_{\alpha \in A} \sigma_\alpha(x)$ is defined and equal to 1. It is *subordinate to the covering* $\mathcal{U} = \{U_\beta \mid \beta \in B\}$ if, for all $\alpha \in A$, there exists $\beta(\alpha) \in B$ such that $\operatorname{supp}(\sigma_\alpha) \subset U_{\beta(\alpha)}$, where $\operatorname{supp}(\sigma_\alpha)$ is the closure of the set of points at which σ_α is non-zero.

In the case of a compact manifold, we may produce partitions of unity by using the $V_p = {\phi_p}^{-1}(B(0, r_p))$ and f_p defined at the beginning of the proof of Theorem 1.4.3. If we cover M by finitely many such open neighbourhoods V_{p_i}, with corresponding bump functions f_{p_i}, then $\sigma = \sum_i f_{p_i}$ is defined, smooth and greater than or equal to 1 everywhere on M. So $\sigma_i = f_{p_i}/\sigma$ is a set of smooth functions on M such that $\sum_i \sigma_i = 1$. It is subordinate to the cover $\{U_i \mid i = 1, \cdots, k\}$ of M. We may make the partition subordinate to a given cover by choosing each of our original coordinate neighbourhoods U_α to lie within a member of that cover.

Many important smooth manifolds, for example Euclidean space $\mathbb{R}^m$ or the total space $\tau(M)$ of the tangent bundle to be introduced below, are non-compact. So it will be important for us to have partitions of unity on all our manifolds and still to find them subordinate to a given cover.

To achieve that we first express the manifold M as a union of closed compact subsets analogous to the manner in which $\mathbb{R}^m$ is the union of the set of closed compact balls $\{\overline{B(0, n)} \mid n \in \mathbb{N}\}$. Choose a countable basis for the topology of M and arrange its members in a sequence. Then, defining sets $U_p \supset W_p \supset V_p$ for each $p \in M$ as in the proof of Theorem 1.4.3, each of the sets V_p is a union of members from the sequence of basic sets. Omitting from the sequence any basic set that is not so required at any point p and, for each remaining basic set, choosing any V_p that contains it, we arrive at a countable cover of M by sets $\{V_{p_n} \mid n \in \mathbb{N}\}$.

Next, we may inductively define an increasing sequence $i(n)$, with $i(1) = 1$, and a re-ordering of the V_{p_n} such that $K_n = W_{p_1} \cup W_{p_2} \cup \cdots \cup W_{p_{i(n)}} \subset V_{p_1} \cup V_{p_2} \cup \cdots \cup V_{p_{i(n+1)}}$. This is possible since K_n is compact: $V_{p_1}, \cdots, V_{p_{i(n)}}$ are already in K_n and $V_{p_{i(n)+1}}, \cdots, V_{p_{i(n+1)}}$ will be the remaining ones that are necessary to cover it. This has the result that $K_n \subset \operatorname{int}(K_{n+1})$, for each n.

Finally, given an open cover $\mathcal{U}$ of M, we apply our preliminary result for compact manifolds to each of the compact subsets $K_n \setminus \operatorname{int}(K_{n-1})$ restricting each of the coordinate neighbourhoods U_p that we use to lie both in the open set $\operatorname{int}(K_{n+1}) \setminus K_n$ and in a member of $\mathcal{U}$. Then a point $x \in K_n$ will only be non-zero under the finitely many bump functions required for K_{n+1}. Thus, as in the compact case, the sum of the bump functions is everywhere defined and we may similarly obtain a partition of unity on M subordinate to the given cover $\mathcal{U}$. Thus we have established the following fundamental result.

Theorem 1.5.2. *Given an open cover $\mathcal{U}$ of a manifold M, there is a partition of unity on M subordinate to $\mathcal{U}$.* ■

Note on paracompactness. What we achieved above for the family of supports of the selected bump functions was that it be *locally finite*: each point $x \in M$ has a neighbourhood ($\text{int}(K_{n+1})$ when $x \in K_n$) meeting the supports of only finitely many of the chosen bump functions. In fact the same remains true for the cover $\mathcal{U}'$ by the open coordinate neighbourhoods that we used to define the bump functions. We also ensured that the cover $\mathcal{U}'$ was a *refinement* of the given open cover $\mathcal{U}$: for every $U' \in \mathcal{U}'$, there exists $U \in \mathcal{U}$ with $U' \subseteq U$. Thus we have, in particular, shown that our manifolds are paracompact:

Definition 1.5.3. The topological space $(X, \mathcal{T})$ is said to be *paracompact* if any covering of X by subsets from $\mathcal{T}$ has a locally finite refinement, by sets from $\mathcal{T}$, that is also a covering.

Provided they are Hausdorff, general paracompact spaces share the property that we have just established for manifolds: if $(X, \mathcal{T})$ is a Hausdorff, paracompact space then, for each open covering $\mathcal{U}$ of X, there exists a subordinate partition of unity.

This relies on the fact that a Hausdorff paracompact space is *normal*, that is, disjoint closed subsets can be separated by disjoint open subsets containing them. We recall that it is not hard to prove this for a compact Hausdorff space. The argument depends on picking a finite subcover for a special covering of X and can be modified to hold under the weaker condition of paracompactness.

Arguing very much as above for manifolds, the result now follows from the Tietze Extension Theorem. This is a classical result that states that, if A and B are disjoint closed subsets of the normal topological space X, then there is a continuous function $f : X \to [0, 1]$ such that $f(A) = 0$ and $f(B) = 1$.

In the case of a smooth manifold, the existence of smooth partitions of unity is important for several existence theorems, some of which we shall meet later in this book, and the partition of unity constructed above will consist of smooth functions if the initial bump functions are smooth. That will be the case, for example, for paracompact manifolds locally homeomorphic to separable Hilbert space. Thus the results that derive from the existence of partitions of unity generalise at least to that context.

1.6. Exercises for Chapter 1.

1.1. Show that the three properties in Definition 1.1.1 are independent of each other.

1.2. Show that the structure of a differentiable manifold which $\mathbf{S}^n$ inherits as a submanifold of $\mathbb{R}^{n+1}$, in Example 1.3.9 (1), coincides with that given in Example 1.1.7 (2).

1.3. Let $\mathcal{A} = \{(U_\alpha, \varphi_\alpha) \mid \alpha \in A\}$ be an atlas on a smooth manifold M^m. Show that a subset U of M is open if and only if $\varphi_\alpha(U \cap U_\alpha)$ is open in $\mathbb{R}^m$ for all $\alpha \in A$.

1.4. Let $\mathbf{S}^2$ have the complex analytic atlas mentioned in Example 1.1.7 (3). Show how any Möbius transformation defined on one of the coordinate spaces may be extended to define a complex analytic homeomorphism of $\mathbf{S}^2$ onto itself.

1.5. Prove that $SO(n)$ is a differentiable submanifold of $\mathbb{R}^{n^2}$.
Prove that $SO(3)$ is diffeomorphic with $\mathbb{RP}^3$.

1.6. A group G acts on a smooth manifold M^m by smooth maps in such a way that each two points p, $q \in M$, with q not in the orbit $G(p)$ of p, have neighbourhoods U, V respectively both of which are disjoint from all the subsets $\{gU \mid g \in G,\, g \neq \mathrm{id}\}$.
Define an atlas on the orbit space M/G making it into a smooth manifold such that the quotient map $M \to M/G$ is a local diffeomorphism.
Find the following actions:
(*i*) of $\mathbb{Z}/2\mathbb{Z}$ on $\mathbf{S}^n$ such that the orbit space is $\mathbb{RP}^n$;
(*ii*) of $\mathbb{Z}$ on $\mathbb{R}$ such that the orbit space is $\mathbf{S}^1$;
(*iii*) of the group $G = \langle g, h \mid ghg = h\rangle$ on $\mathbb{R}^2$ such that the orbit space is a Klein bottle.

1.7. Show that the image of $\mathbf{S}^2 \subset \mathbb{R}^3$ under the map

$$f : (x, y, z) \mapsto (x^2, y^2, z^2, \sqrt{2}\, yz, \sqrt{2}\, zx, \sqrt{2}\, xy)$$

is a submanifold of $\mathbb{R}^6$. How does it relate to $\mathbb{RP}^2$?

1.8. Show that the composition of two embeddings is an embedding and that the product of two embeddings is an embedding.
If M^m is a product of spheres, prove that it can be embedded in $\mathbb{R}^{m+1}$.

CHAPTER 2

THE DERIVATIVES OF DIFFERENTIABLE MAPS

In the first chapter we defined differential manifolds and differentiable maps between them. However, unlike the Euclidean case which motivated it, these differentiable maps had no derivatives. We now remedy that situation with a discussion of tangents, cotangents and the maps between them that generalise the derivatives of maps between Euclidean spaces. We start with the local picture, the view through a chart.

2.1. Tangent Vectors.

Recall that, for $U \subseteq \mathbb{R}^m$, $f : U \to \mathbb{R}^n$ is differentiable at p if there exists a linear map $Df(p)$ such that

$$f(p+v) - f(p) = Df(p) \cdot v + o(\|v\|).$$

The domain of $Df(p)$ is $\mathbb{R}^m$ identified as 'displacement vectors' v at p, which we may write as v_p. Note that the assumption that $Df(p)$ is linear means that, for t real, $Df(p) \cdot (tv) = tDf(p) \cdot v$, so that, fixing p and v, $f(p+tv) - f(p) = t\,Df(p) \cdot v + o(\|t\|)$. Thus, if γ is the curve $\gamma : t \mapsto p + tv$,

$$Df(p) \cdot v = \lim_{t \to 0} \frac{f(p+tv) - f(p)}{t} = \lim_{t \to 0} \frac{f(\gamma(t)) - f(\gamma(0))}{t} = (f \circ \gamma)'(0).$$

This is $Df(\gamma(0)) \cdot \gamma'(0)$ by the chain rule, unsurprisingly since $\gamma(0) = p$ and $\gamma'(0) = v$. However, it is also $Df(\delta(0)) \cdot \delta'(0)$ for any other curve δ such that $\delta(0) = p$ and $\delta'(0) = v$. So, for any $v \in \mathbb{R}^m$, we may identify the vector v_p, 'v acting at p', with the equivalence class $[\gamma]_p$ of curves γ with $\gamma(0) = p$, where $\gamma \sim \delta \Leftrightarrow \gamma'(0) = \delta'(0) = v$. Unlike our original way of looking at vectors in Euclidean space, this revised point of view works for manifolds:

Definition 2.1.1. A *tangent vector at p in M* is an equivalence class $[\gamma]$ of (smooth) curves $\gamma : \mathbb{R} \to M$ such that $\gamma(0) = p$ and $\gamma \sim \delta$ if, for some chart ϕ at p, $(\phi \circ \gamma)'(0) = (\phi \circ \delta)'(0)$.

Remarks 2.1.2. 1) This is well-defined, that is, it is independent of the chart ϕ.

2) We write simply v for $[\gamma]$ when no representative curve is specified. We also write v_p or $[\gamma]_p$ when we wish to specify the point at which the vector acts.

3) The definition agrees with the example that motivated it since, using the identity chart on $\mathbb{R}^n$, curves γ and δ in $\mathbb{R}^n$ with $\gamma(0) = \delta(0)$ are equivalent if and only if $\gamma'(0) = \delta'(0)$. Then if, as before, $\gamma(0) = p$ and $\gamma'(0) = v$ we have $[\gamma]_p = v_p$.

4) We only need γ, δ defined on a neighbourhood of 0 in $\mathbb{R}$.

5) There are other possible (equivalent) definitions of tangent vectors, one of which we shall introduce later in this chapter and another of which we shall introduce in Chapter 3 as an example of a general method for constructing fibre bundles.

Definition 2.1.3. If $f : M \to N$ is differentiable, then its *derivative at p* is the map

$$f_*(p) : [\gamma] \mapsto [f \circ \gamma]_{f(p)}.$$

Remarks 2.1.4. 1) We could denote this by $Df(p)$, but we'll reserve that notation for the derivative of a map $f : U \to \mathbb{R}^n$ where $U \subseteq \mathbb{R}^m$.

2) For a further differentiable map $g : N \to P$, $(g \circ f)_*(p) = g_*(f(p)) \circ f_*(p)$, the analogue of the chain rule. Also $(\mathrm{id})_* = \mathrm{id}$.

3) Note that, for a curve γ in $\mathbb{R}^m$, strictly $\gamma'(0)$ is $\gamma_*(0) \cdot 1$, which is indeed $[\gamma]_{\gamma(0)}$ since the unit tangent vector 1 to the real line is the class of the identity map. Similarly, if τ_t is the translation $s \mapsto s+t$, then $\gamma'(t) = \gamma_*(t) \cdot 1$ is the class $[\gamma \circ \tau_t]_{\gamma(t)}$. It will be convenient to use the notation $\gamma'(t)$ for $\gamma_*(t) \cdot 1 = [\gamma \circ \tau_t]_{\gamma(t)}$ also when it is a tangent vector on a manifold.

4) For $U \subseteq \mathbb{R}^m$ and $f : U \to \mathbb{R}^n$ our initial discussion showed that $f_*(p)$ is just $Df(p)$ when each $[\gamma]_p$ is identified with $\gamma'(0)$.

Proposition 2.1.5. *The derivative $f_*(p)$ is well-defined.*

Proof. Suppose $\gamma \sim \delta$. Then $(\phi \circ \gamma)'(0) = (\phi \circ \delta)'(0)$ for a chart (ϕ, U) at p. So, for a chart (ψ, V) at $f(p)$,

$$\begin{aligned}(\psi \circ f \circ \gamma)'(0) &= D(\psi \circ f \circ \phi^{-1})\,(\phi(\gamma(0))) \circ (\phi \circ \gamma)'(0) \\ &= D(\psi \circ f \circ \phi^{-1})\,(\phi(\delta(0))) \circ (\phi \circ \delta)'(0) \\ &= (\psi \circ f \circ \delta)'(0)\end{aligned}$$

by the chain rule, since $\psi \circ f \circ \phi^{-1}$ is differentiable. Thus $f \circ \gamma \sim f \circ \delta$ as required. ∎

2.2. The Tangent Space.

Given a chart (ϕ, U) at p, with $\phi(p) = 0 \in V \subseteq \mathbb{R}^m$, the map $\gamma \mapsto \phi \circ \gamma$ is a bijection between (smooth) curves γ in U such that $\gamma(0) = p$ and smooth curves $\tilde{\gamma} = \phi \circ \gamma$ in V such that $\tilde{\gamma}(0) = 0$. See Figure 2.1.

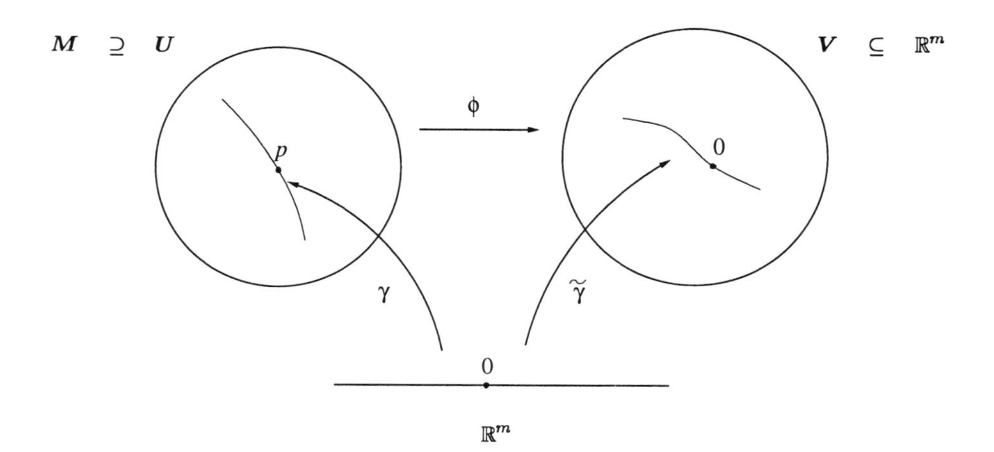

Figure 2.1.

We also see that

$$\gamma \sim \delta \iff (\phi \circ \gamma)'(0) = (\phi \circ \delta)'(0) \iff \phi \circ \gamma \sim \phi \circ \delta$$

by Remark 2.1.4 (2). So we have a bijection

$$\tau_p(M) = \{\text{tangent vectors } [\gamma]_p \text{ at } p\} \longleftrightarrow \{\text{tangent vectors at } 0 \text{ on } \mathbb{R}^m\}.$$

However the set of tangent vectors v_0 at 0 in $\mathbb{R}^m$ is bijective with $\mathbb{R}^m$ itself and so has a natural vector space structure. The above bijection then transfers this to $\tau_p(M)$. In other words, if $[\gamma]_p$ corresponds to v_0 and $[\delta]_p$ corresponds to w_0 then $\lambda[\gamma]_p + \mu[\delta]_p$ is the tangent vector at p which corresponds to $(\lambda v + \mu w)_0$. In fact, since we have chosen ϕ such that $\phi(p) = 0$, it is the vector $[\phi^{-1}(\lambda\phi \circ \gamma(t) + \mu\phi \circ \delta(t))]_p$. As usual this is well defined since the linear structures induced by different charts differ by the linear isomorphism $D\theta$ where θ is the coordinate transformation. Note however that, unlike in $\mathbb{R}^m$ itself, there is no longer a standard choice of basis.

Proposition 2.2.1. *For a (smooth) map $f : M^m \to N^n$ the derivative $f_*(p) : \tau_p(M) \to \tau_{f(p)}(N)$ is linear.*

Proof. Take charts (ϕ, U) at p and (ψ, V) at $f(p)$. Then the linear structure of $\tau_p(M)$ is determined by $\phi_*(p)$ from that of $\tau_0(\mathbb{R}^m) \cong \mathbb{R}^m$ and the linear structure of $\tau_{f(p)}(N)$ is determined similarly by $\psi_*(f(p))$ from that of $\tau_0(\mathbb{R}^m)$. But $D(\psi \circ f \circ \phi^{-1})$ maps $\tau_0(\mathbb{R}^m)$ linearly onto $\tau_0(\mathbb{R}^n)$. See Diagram 2.2, in which the lower block indicates vectors and the maps between them and the upper block depicts the various spaces to which those vectors are tangent and the maps between those spaces. ∎

Special case. On $\mathbb{R}^m$ each tangent space $\tau_p(\mathbb{R}^m) = \{v_p \mid v \in \mathbb{R}^m\}$ is naturally identified with $\mathbb{R}^m$ itself. The resulting identification of $\tau_p(\mathbb{R}^m)$ with $\tau_q(\mathbb{R}^m)$ is referred to as *parallel translation.* Indeed the identification of $\tau_p(\mathbb{R}^m)$ with $\tau_q(\mathbb{R}^m)$ is just the derivative of the translation mapping $x \mapsto x + q - p$ of $\mathbb{R}^m$ to itself that takes p to q.

So for a smooth map $f : M^m \to \mathbb{R}$ we have the composition

$$\tau_p(M^m) \xrightarrow{f_*(p)} \tau_{f(p)}(\mathbb{R}) \xrightarrow{\cong} \mathbb{R}$$

called $df(p)$ and referred to as the *differential* of f. This lies in the dual space $\mathcal{L}(\tau_p(M), \mathbb{R})$ which we denote by $\tau_p^*(M)$. These elements are called *covectors* or *cotangent vectors* and $\tau_p^*(M)$ is called the *cotangent space.*

2.3. The Tangent and Cotangent Bundles.

We now bring the derivative into the category of differential manifolds and maps by producing a manifold on which a global derivative, the union of all the derivatives defined on the tangent spaces, may act.

$$\begin{array}{ccccccc}
 & & \mathbb{R} & & & & \\
 & & \downarrow{\scriptstyle \gamma} & & & & \\
M & \supseteq & U & \xrightarrow{f} & V & \subseteq & N \\
 & & \downarrow{\scriptstyle \phi} & & \downarrow{\scriptstyle \psi} & & \\
\mathbb{R}^m & \supseteq & \phi(U) & \xrightarrow{\psi \circ f \circ \phi^{-1}} & \psi(V) & \subseteq & \mathbb{R}^n
\end{array}$$

$$\begin{array}{ccccccc}
\tau_p(M) & \ni & [\gamma]_p & \xrightarrow{f_*(p)} & [f \circ \gamma]_{f(p)} & \in & \tau_{f(p)}(N) \\
 & & \downarrow{\scriptstyle \phi_*(p)} & & \downarrow{\scriptstyle \psi_*(f(p))} & & \\
\tau_0(\mathbb{R}^m) & \ni & [\phi \circ \gamma]_{\phi(p)} & \xrightarrow{\tilde{f}(p)} & [\psi \circ f \circ \gamma]_{\psi(f(p))} & \in & \tau_0(\mathbb{R}^n)
\end{array}$$

The map $\tilde{f}(p)$ is $(\psi \circ f \circ \phi^{-1})_*(\phi(p)) \equiv D(\psi \circ f \circ \phi^{-1})(\phi(p))$.

Diagram 2.2.

For any open subset U of M, possibly M itself and not necessarily a coordinate neighbourhood, we write $\tau(U)$ for the, intially disjoint, union $\bigcup_{p \in U} \tau_p(M)$ of all the tangent spaces at points of U. We then introduce topological and differential structures on $\tau(U)$ in three stages.

(*a*) For $V \subseteq \mathbb{R}^m$ we have $\tau(V) \cong V \times \mathbb{R}^m$ using the parallel translation isomorphisms at each point. We take this differential structure on $\tau(V)$, regarding $V \times \mathbb{R}^m$ as a subset of $\mathbb{R}^{2m}$.

(*b*) For a chart (ϕ, U) on M we have the bijection

$$\Phi : \tau(U) \longrightarrow \tau(\phi(U)); \quad [\gamma]_x \overset{\phi_*(x)}{\longmapsto} [\phi \circ \gamma]_{\phi(x)} \,,$$

which is linear on each tangent space $\tau_x(U)$. We take the topological and differential structures which this bijection induces on $\tau(U)$.

(*c*) Different charts on (part of) U induce structures (topological and differential) differing by the diffeomorphism

$$\Theta_{21} : V_1 \times \mathbb{R}^m \longrightarrow V_2 \times \mathbb{R}^m; \quad (x, v) \mapsto (\theta_{21}(x), D\theta_{21}(x) \cdot v).$$

Then θ_{21} smooth implies that $D\theta_{21}$ is smooth and hence so is Θ_{21}. Thus the topologies induced are the same and the charts Φ_i on $\tau(U_i)$ are compatible.

We check that $\tau(M)$ is

(*i*) second countable: take a countable cover of M by coordinate neighbourhoods with coordinate spaces V_i; then each $V_i \times \mathbb{R}^m$ is second countable and so too are the $\tau(U_i)$ which form an open covering of $\tau(M)$.

(*ii*) Hausdorff: if $v_p \neq w_q$ then either $p \neq q$ and we may use the Hausdorff property of M, or $p = q$ and $w \neq v$ and we use the Hausdorff property of $\mathbb{R}^m$.

Proposition 2.3.1. *The projection* $\pi : \tau(M) \to M$; $v_p \mapsto p$ *is a differentiable submersion.* ■

Definition 2.3.2. The set $\tau(M)$ with the above topological and differential structure is called the *tangent bundle* of M.

Fibre bundles in general and vector bundles in particular, of which the tangent bundle is but one example, will be discussed in more detail in Chapter 3. As a set the tangent bundle $\tau(M)$ is easy to visualise:

$$\tau(M) = \bigcup_{p \in M} \tau_p(M).$$

The essential point to note about its topology is that it suffices to define this *locally*, that is, to describe a system of small neighbourhoods for each point in the space. This we can do, since $\tau(M)\big|_U \cong U \times \mathbb{R}^m$ for a family of open subsets U covering M. It may help to use the Whitney Embedding Theorem from Section 1.4 and, at least in the compact case, to embed M^m in $\mathbb{R}^n$ for sufficiently large n. Since, as a smooth manifold, $\mathbb{R}^n$ has a single coordinate neighbourhood, $\tau(\mathbb{R}^n) \cong \mathbb{R}^n \times \mathbb{R}^n$, and the local definition of the topology on $\tau(M)$ shows that

$$\tau(M) \hookrightarrow \mathbb{R}^{2n}$$

as a submanifold. The reader may like to consider, and indeed attempt to draw a picture for, the following special case. The sphere $\mathbf{S}^2$ is defined as a submanifold of $\mathbb{R}^3$ (Example 1.3.9 (1)) and hence

$$\tau(\mathbf{S}^2) \hookrightarrow \mathbb{R}^6.$$

The 3-dimensional submanifold of $\tau(\mathbf{S}^2)$ formed by the tangent vectors of unit length in fact embeds in $\mathbb{R}^5$ (Wall, 1965) but, for homological reasons not in $\mathbb{R}^4$. So the embedding above for the full tangent bundle is actually very economical.

The tangent bundle is the natural domain for the derivative of a differentiable map defined on M:

Proposition 2.3.3. *If $f : M^m \to N^n$ is differentiable, so is $f_* : \tau(M) \to \tau(N)$; $v_p \mapsto (f_*(p))(v)$.*

Proof. Locally charts (ϕ, U) on M and (ψ, V) on N give charts $(\Phi, \tau(U))$ on $\tau(M)$ and $(\Psi, \tau(V))$ on $\tau(N)$. Then, since Φ restricted to $\tau_p(M)$ is just $\phi_*(p)$ and $(\phi_*(p))^{-1} = (\phi^{-1})_*(\phi(p))$, for $w = \Phi(v_p) = \phi_*(p) \cdot v_p$ we have

$$\begin{aligned}
\Psi \circ f_* \circ \Phi^{-1}(w) &= \psi_*(f(p)) \circ f_*(p) \circ (\phi_*(p))^{-1}(w) \\
&= \psi_*(f(p)) \circ f_*(p) \circ (\phi^{-1})_*(\phi(p))(w) \\
&= (\psi \circ f \circ \phi^{-1})_*(\phi(p))(w) \\
&= D(\psi \circ f \circ \phi^{-1})(\phi(p))(w).
\end{aligned}$$

Thus $\Psi \circ f_* \circ \Phi^{-1}$ is differentiable and so, by definition, is f_*. ■

Having used the bijection $\phi_*(p) : \tau_p(M) \to \tau_{\phi(p)}(\mathbb{R}^m)$ derived from the chart $\phi : U \to V \subseteq \mathbb{R}^m$, to induce a vector space structure on $\tau_p(M)$, we have the dual isomorphism $\phi^*(p)(= (\phi_*(p))^*) : \tau_p^*(M) \leftarrow \tau_{\phi(p)}^*(\mathbb{R}^m)$ and, letting p vary in U, the bijection, linear on each cotangent space $\phi^* : \tau^*(U) \leftarrow \tau^*(\phi(U))$. More naturally, using $(\phi^*(p))^{-1} = (\phi^{-1})^*(\phi(p))$, we have

$$(\phi^{-1})^* : \tau^*(U) \to \tau^*(\phi(U)).$$

Now, if $\theta : U \to V$ is a diffeomorphism between open subsets of $\mathbb{R}^m$ and $D\theta(x)$ has matrix $J\theta(x)$ with respect to the standard basis, then $(D\theta^{-1})(\theta(x))^*$ has matrix $((J\theta(x))^{-1})^T$ with respect to the dual bases and, in particular, it is an isomorphism.

Thus we may use the $(\phi^{-1})^*$ to define the structure of the *cotangent bundle* $\tau^*(M)$, just as we used ϕ_* to define that of the tangent bundle, the necessary coordinate transformations being derived from those of the manifold as before. Indeed, as we shall see in Chapter 3, a whole range of bundles 'associated with the tangent bundle' may be similarly defined. Although these bundles are important in themselves, it is more often through their cross-sections that we get to know and understand them: a cross-section, or just *section* of a bundle projection $\pi : E \to M$ is a map $\sigma : M \to E$ such that $\pi \circ \sigma = \mathrm{id}_M$. These sections are usually sufficiently important to acquire special names of their own, and we now name the two most basic sections.

Definition 2.3.4. A smooth map $X : M \to \tau(M)$ such that $\pi \circ X = \mathrm{id}_M$ is called a *vector field on* M. A map $\omega : M \to \tau^*(M)$ such that $\pi \circ \omega = \mathrm{id}_M$ is called a *covector field* or 1-*form on* M.

Remarks 2.3.5. 1) For a vector field X, $\pi \circ X = \mathrm{id}_M$ implies that $X(p) \in \tau_p(M)$ for all $p \in M$. Thus X picks out a tangent vector at each point of M. But it does so 'smoothly' in the sense made precise above.

The reader may be familiar with the common representation of a vector field X in the plane, or a subset of it, under which, for a suitable selection of points p, the vector $X(p)$ is indicated by an arrow emanating from p whose magnitude and direction is determined by $X(p)$. For example, Figure 2.3 shows a chart indicating the forecast wind speeds and directions over a piecewise linear approximation to a familiar group of islands. In this case the domain is not a plane but a subset, containing those islands, of the surface of the earth which, to a first approximation, is diffeomorphic with the 2-sphere.

2) If $f : M \hookrightarrow N$ is an embedding, then $f_*(\tau_p(M))$ is a subspace of $\tau_{f(p)}(N)$ which can be regarded as $\tau_{f(p)}(M)$ on identifying M with $f(M)$.

Example 2.3.6. For $\mathbf{S}^m \hookrightarrow \mathbb{R}^{m+1}$ as unit sphere the tangent space at $x \in \mathbf{S}^m$ 'is'

$$\{v_x \mid v \cdot x = 0\} \subseteq \tau_x(\mathbb{R}^{m+1}).$$

For, if $\iota : \mathbf{S}^m \hookrightarrow \mathbb{R}^{m+1}$ is the inclusion and $\gamma : \mathbb{R} \to \mathbf{S}^m$ such that $\gamma(0) = x$, then $[\iota \circ \gamma]_x = (\iota \circ \gamma)'(0)$ is perpendicular to x.

Example 2.3.7. For $v \in \mathbb{R}^2$, let

$$Av = \begin{pmatrix} 0 & -1 \\ 1 & 0 \end{pmatrix} v$$

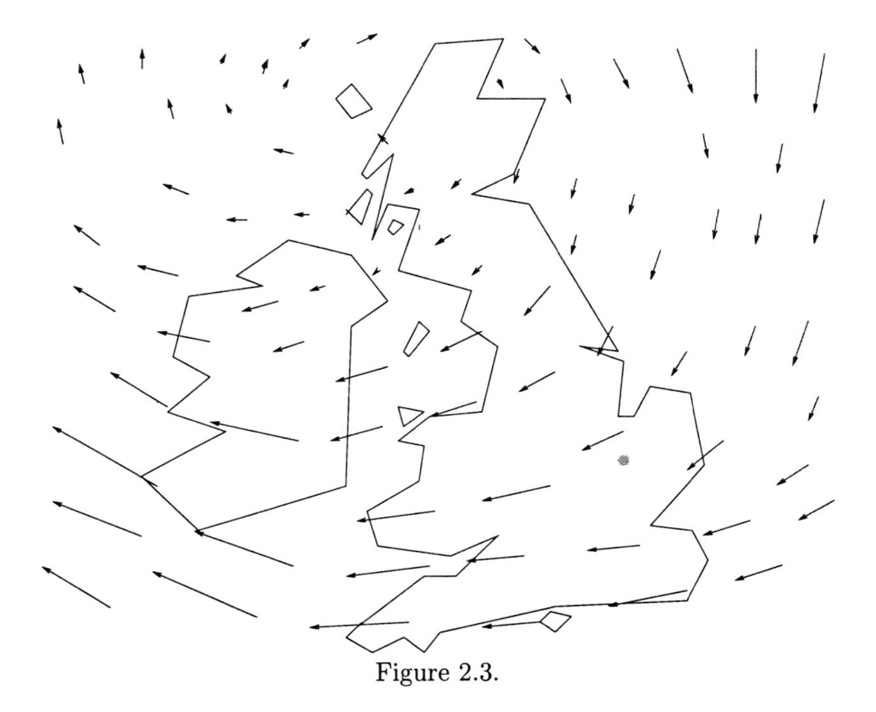

Figure 2.3.

be v rotated through an angle $\pi/2$. Then

$$X : \mathbf{S}^1 \longrightarrow \tau(\mathbf{S}^1) \hookrightarrow \tau(\mathbb{R}^2)\big|_{\mathbf{S}^1}; \quad v \mapsto (Av)_v$$

is a never zero vector field on $\mathbf{S}^1$, five vectors of which are illustrated in Figure 2.4.

2.4 Whitney's Embedding Theorem Revisited.

Now that we have defined the tangent bundle we are in a position to establish a stronger version of Whitney's Embedding Theorem, although the reader may find it helpful to refer to the relevant parts of section 3.2 at times. As with all of Chapter 3, this section may be omitted at a first reading.

Theorem 2.4.1. (Whitney.) A compact manifold M^m of dimension m may be immersed in $\mathbb{R}^{2m}$ and embedded in $\mathbb{R}^{2m+1}$.

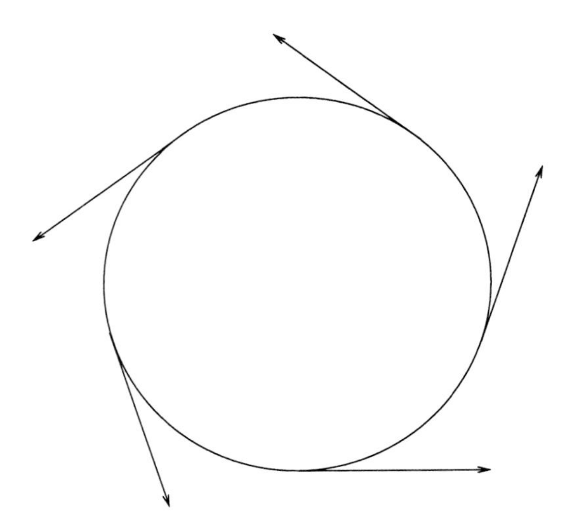

Figure 2.4.

Proof. Assume that the compact manifold M^m is embedded in $\mathbb{R}^n$, which we denote by $M \hookrightarrow \mathbb{R}^n$, and let $\mathbb{R}^{n-1} = \{x \in \mathbb{R}^n \big| x_n = 0\}$. For each unit vector v in $\mathbb{R}^n \setminus \mathbb{R}^{n-1}$ we consider the projection, π_v, of M into $\mathbb{R}^{n-1}$ parallel to v. We claim that, provided $n > 2m+1$, it is possible to choose v so that π_v is an injective immersion and hence, since M is compact, an embedding. Repeating this $n - 2m - 1$ times, we arrive at an embedding of M in $\mathbb{R}^{2m+1} = \{x \in \mathbb{R}^n \big| x_{2m+2} = 0, \ldots, x_n = 0\}$.

To ensure that π_v is an immersion we need $\frac{z}{\|z\|} \neq v$ for each tangent vector z to M, since then $(\pi_v)_*(z) \neq 0$. Clearly it suffices to consider the unit vectors z, which form a submanifold $\sigma(M)$ of $\tau(M)$. Let $f : \sigma(M) \longrightarrow \mathbf{S}^{n-1}$ be the parallel translation to the sphere of unit tangent vectors at the origin. Since the tangent bundle to $\mathbb{R}^m$ is trivial, the map f may be thought of as the composition of the inclusion in $M \times \mathbb{R}^n$ followed by projection on the second factor. So f is smooth and, by Sard's theorem, since $\dim(\sigma(M)) = 2m-1 < n-1$, $f(\sigma(M)) \subset \mathbf{S}^{n-1}$ has measure zero. Since an open subset cannot have measure zero, every open subset of $\mathbf{S}^{n-1}$ must meet the complement of $f(\sigma(M))$ and so contain vectors v such that π_v is an immersion. Note that this stage of our proof only required $2m < n$ so that, by repeating it, we can obtain an immersion of M in $\mathbb{R}^{2m}$.

To ensure that π_v is injective we require that, for any two points x, y in

$M \subset \mathbb{R}^n$, the vector $y - x$ is not parallel to v. So, denoting by Δ the set $\{(x, x)\} \subset M \times M$, we consider the map

$$g : M \times M \setminus \Delta \longrightarrow \mathbf{S}^{n-1}; (x, y) \mapsto \frac{y - x}{\|y - x\|}.$$

The domain of g has dimension $2m$ so again, provided $2m < n - 1$, the image has measure zero. However, since the domain of f is compact, its image is also compact and so closed. Thus the complement of $f(\sigma(M))$ in $\mathbf{S}^{n-1}$ is a non-empty open set which must therefore contain points v not in the image of g. As already explained, for such v, π_v will be an embedding and this argument can be repeated to obtain the required embedding of M in $\mathbb{R}^{2m+1}$. ■

2.5. Vector Fields and 1-forms on Euclidean Spaces.

Although vector fields and 1-forms are global concepts, when performing explicit calculations we shall almost always work locally. That is, we shall use coordinate charts to transform them into fields and forms on the coordinate spaces, subspaces of Euclidean spaces. Hence it is important to see what these concepts look like in that context. In doing so we shall also see how our definitions have generalised the more familiar Euclidean calculus.

Example 2.5.1. Consider the paths

$$\delta_i : \mathbb{R} \longrightarrow \mathbb{R}^m; \quad t \mapsto (x_1, \cdots, x_{i-1}, x_i + t, x_{i+1}, \cdots, x_m)$$

and the projections

$$\pi_i : \mathbb{R}^m \longrightarrow \mathbb{R}; \quad (x_1, \cdots, x_m) \mapsto x_i.$$

Then $[\delta_i]_x = \delta_i'(0) = e_i$, the ith standard basis vector of $\mathbb{R}^m$ translated to act at x. Thus the vectors $\{[\delta_i]_x \mid i = 1, \cdots, m\}$ form a basis of $\tau_x(\mathbb{R}^m)$. However, recalling the definition of the differential $d\pi_i$ at the end of Section 2.2,

$$d\pi_i(x)([\delta_j]_x) = [\pi_i \circ \delta_j]_{x_i} = (\pi_i \circ \delta_j)'(0) = \delta_{ij}.$$

Here δ_{ij}, not to be confused with the path δ_i, is the Kronecker delta, which takes the value 1 if the suffixes i and j are equal and otherwise is zero. So

$\{d\pi_i(x) \mid i = 1, \cdots, m\}$ is the basis of $\tau_x^*(\mathbb{R}^m)$ dual to the basis $\{[\delta_i]_x \mid i = 1, \cdots, m\}$ of $\tau_x(\mathbb{R}^m)$.

An alternative definition of tangent vectors will give us further insight into these basis vectors. Recall that, fixing $f : U \to \mathbb{R}$, for U open in $\mathbb{R}^m$, we defined the element $df(p)$ of the cotangent space $\tau_p^*(U) = \mathcal{L}(\tau_p(U), \mathbb{R})$ by $df(p)(v_p) = [f \circ \gamma]_{f(p)}$ where $v_p = [\gamma]_p$, or equivalently $df(p)(v_p) = (f \circ \gamma)'(0) \in \tau_{f(p)}(\mathbb{R})$, followed by parallel translation of $\tau_{f(p)}(\mathbb{R})$ onto $\mathbb{R}$ itself. If, instead, we fix v_p and let f vary in $\mathcal{C}^\infty(U)$, the algebra of smooth functions on U, then $v_p = [\gamma]_p$ defines the linear map

$$v_p : \mathcal{C}^\infty(U) \longrightarrow \mathbb{R}; \quad f \mapsto (f \circ \gamma)'(0).$$

This satisfies

$$v_p(fg) = f(p)\, v_p(g) + g(p)\, v_p(f) \tag{2.1}$$

and may be called a *derivation at p of the algebra* $\mathcal{C}^\infty(U)$. We leave it as an exercise, Exercise 2.1, at the end of this chapter for the reader to check that this is equivalent to our first definition of tangent vectors.

Having introduced yet another symbol for the same real number it is perhaps worth summarising them. For a function f on an open set U, in $\mathbb{R}^m$ or indeed on a manifold, and a tangent vector v_p at $p \in U$, all the following are equal:

(*i*) $v_p(f)$, the value of the derivation v_p at p on the local function f;

(*ii*) $df(p)(v_p)$, the value of the differential of f at p on the vector v_p, which, as v varies, expresses $df(p)$ as a covector at p;

(*iii*) $f_*(v_p)$, the image of v_p under the derivative of f;

(*iv*) $[f \circ \gamma]_{f(p)} = f_*([\gamma]_p)$, the equivalence class of curves representing $f_*(v_p)$ when the class $[\gamma]_p$ represents v_p;

(*v*) $(f \circ \gamma)'(0)$, the classical notation for the derivative of the real-valued function $f \circ \gamma$ of a real variable.

Strictly speaking, (*iii*) and (*iv*) should be thought of as tangent vectors to $\mathbb{R}$ at $f(p)$ and the other three as real numbers, but modulo parallel translation they are the same.

We urge the reader not to become overwhelmed by this plethora of definitions. It is precisely the fact that there are so many different ways of looking at essentially the same object that leads to the beauty and power of differential geometry and analysis. For example, allowing p to vary in (*i*) will give us an alternative definition of a vector field that we shall exploit in the next section.

Alternatively, taking v_p to be a unit vector in (*ii*) allows us to interpret the common value as the directional derivative of f at p in the direction v_p.

Returning to the basis vectors of $\tau_x(\mathbb{R}^m)$, when the vector $[\delta_j]_x$ is regarded as a derivation, it is just

$$f \mapsto (f \circ \delta_j)'(0) = \frac{\partial f}{\partial x_j}(x).$$

So we write $\frac{\partial}{\partial x_j}(x)$ or just $\partial_j(x)$ for $[\delta_j]_x$. It is also common to write, illogically but suggestively, $dx_i(x)$ for $d\pi_i(x)$. Then, as we let x vary over a domain V in $\mathbb{R}^m$, we have the vector fields $\partial_j : V \to \tau(V)$ and 1-forms $dx_j : V \to \tau^*(V)$ giving respectively, at each point, the standard basis for $\tau_x(V)$ and the dual basis for $\tau_x^*(V)$. Note that a general vector field X on V may be expressed as $X(x) = a_i(x)\,\partial_i(x)$ with $a_i \in \mathcal{C}^\infty(V)$. Then, by the linearity of f_* in (*iii*),

$$X(x)f = f_*(X(x)) = a_i(x)\, f_*(\partial_i(x)) = a_i(x)\frac{\partial f}{\partial x_i}(x) \tag{2.2}$$

But what is $df(x)$ in terms of the basis $d\pi_i(x)$? Well, if we write $df(x)$ as the linear combination $a_i(x)\,d\pi_i(x)$ of the basis covectors at x, then

$$\begin{aligned}\frac{\partial f}{\partial x_j}(x) &= (f \circ \delta_j)'(0) = df(x)([\delta_j]_x)\\ &= a_i(x)\,d\pi_i(x)([\delta_j]_x) = a_i(x)\delta_{ij} = a_j(x).\end{aligned}$$

So

$$df(x) = \frac{\partial f}{\partial x_i}(x)\,d\pi_i(x) = \frac{\partial f}{\partial x_i}(x)\,dx_i(x)$$

or, allowing x to vary over V,

$$df = \frac{\partial f}{\partial x_i}dx_i \quad !!$$

Thus, once the differentials are interpreted as covector fields, we are able to prove this formula, which you may have first met as a somewhat mysterious definition.

Note that, in the previous paragraph, we used the **summation convention**: if, in an expression, a suffix that can take values in a given finite range is repeated, then it is assumed to be summed over that range. Thus, here $a_i(x)\,d\pi_i(x)$ means $\sum_{i=1}^{m} a_i(x)\,d\pi_i(x)$. We shall maintain this convention throughout the book.

Example 2.5.2. We also compute $\phi_*(\partial_k(x))$ for the differentiable map $\phi : \mathbb{R}^m \longrightarrow \mathbb{R}^n$, for which

$$\phi(x_1, \cdots, x_m) = (y_1, \cdots, y_n) = (\phi_1(x_1, \cdots, x_m), \cdots, \phi_n(x_1, \cdots, x_m)).$$

Then

$$\phi_* \left(\frac{\partial}{\partial x_k}(x) \right) = \phi_*(x)([\delta_k]_x) = [\phi \circ \delta_k]_{\phi(x)} = (\phi \circ \delta_k)'(0) \in \tau_{\phi(x)}(\mathbb{R}^n).$$

But, as we are mapping to a Euclidean space from a Euclidean space, this is the classical derivative

$$D(\phi \circ \delta_k)(0) = D\phi(x) \cdot \delta_k'(0) = D\phi(x) \cdot e_k,$$

since $\delta_k'(0)$ is the standard basis vector e_k acting at $\delta_k(0) = x$. However $D\phi(x)$ is given by the Jacobian matrix $J\phi(x) = \left(\frac{\partial \phi_i}{\partial x_j}(x)\right)$ with respect to the standard bases, so $\phi_*(\partial_k(x)) = \frac{\partial \phi_i}{\partial x_k}(x)\, e_i$, the vector determined by the kth column of $J\phi(x)$, which, still using the summation convention, we may also write as $\frac{\partial \phi_i}{\partial x_k}(x)\, \frac{\partial}{\partial y_i}(\phi(x))$. Thus, for the vector fields, we have

$$\phi_*(\partial_k) = \frac{\partial \phi_i}{\partial x_k}(\partial_i \circ \phi).$$

Classically the '$\circ\phi$' is implied by writing $\phi_* \left(\frac{\partial}{\partial x_k}\right) = \frac{\partial \phi_i}{\partial x_k} \frac{\partial}{\partial y_i}$, it being taken for granted that the x_k are the coordinates on the domain of ϕ and the y_i are the coordinates on the image.

Indeed even more detail is suppressed in practice. For example, if ϕ is a diffeomorphism of $\mathbb{R}^m$, regarded as a change of coordinate system, and $f : \mathbb{R}^m \to \mathbb{R}$ a function defined in terms of the 'y'-coordinates, then $F = f \circ \phi$ is the same function defined in terms of the 'x'-coordinates and

$$\frac{\partial F}{\partial x_k} = \frac{\partial (f \circ \phi)}{\partial x_k} = (f \circ \phi \circ \partial_k)'(0) = \left(\phi_* \frac{\partial}{\partial x_k}\right) f = \frac{\partial \phi_i}{\partial x_k} \frac{\partial f}{\partial y_i}.$$

Since $\phi_i(x_1, \cdots, x_m)$ determines the ith 'y'-coordinate and since one must use ϕ to re-express f in terms of the 'x'-coordinates, it is common to re-write this equation as

$$\frac{\partial f}{\partial x_k} = \frac{\partial y_i}{\partial x_k} \frac{\partial f}{\partial y_i}.$$

Later, in Chapter 4, we shall have reason to denote $f \circ \phi$ by $\phi^*(f)$. Then an alternative way to write the equation would be

$$\frac{\partial}{\partial x_k}(\phi^*(f)) = \left(\phi_* \left(\frac{\partial}{\partial x_k}\right)\right) f.$$

2.6. The Lie Bracket of Vector Fields.

The definition, that we introduced in the previous section, of a tangent vector at p as a derivation at p is not a derivation of the algebra $\mathcal{C}^\infty(U)$ of smooth functions defined on a neighbourhood U of p in the sense defined in Chapter 9. For that we need to consider instead the algebra of *germs of smooth functions at* p. These are equivalence classes of functions defined on (variable) neighbourhoods of p where two functions are equivalent if they agree on some (smaller) neighbourhood of p. This complication is avoided when we pass to the corresponding definition for vector fields: if X is a vector field and f a function on M, then we may define a new function Xf by $(Xf)(p) = X_p(f)$, the value of f under the derivation X_p at p. It follows immediately from the local derivation property (2.1) that

$$X(f\,g) = f\,X(g) + g\,X(f),$$

where the products $f\,g$, etc., are defined pointwise: $(f\,g)(x) = f(x)\,g(x)$. Thus, the vector field X is a derivation of the algebra $\mathcal{C}^\infty(M)$ of smooth (real valued) functions on the manifold. This point of view leads to an easy definition of the Lie bracket of two vector fields, a construction that we shall have occasion to use in later chapters.

Definition 2.6.1. Let X and Y be two vector fields on the smooth manifold M. Then their *Lie bracket* is the vector field defined by

$$[X, Y] = XY - YX,$$

where the composition on the right is the composition of derivations.

That this bracket does define a derivation and hence a new vector field can either be checked directly, see Exercise 5.4, or deduced from the local definition given below. We note that, if U is an open subset of M, then $[X, Y]\big|_U = [X\big|_U, Y\big|_U]$.

Lemma 2.6.2. *The Lie bracket has the following properties for vector fields* X, Y *and* Z *and real scalars* λ:

(*i*) $[X, Y + Z] = [X, Y] + [X, Z]$ *and* $[X, \lambda Y] = \lambda[X, Y]$;
(*ii*) $[X, Y] = -[Y, X]$;
(*iii*) (Jacobi Identity) $[X, [Y, Z]] + [Y, [Z, X]] + [Z, [X, Y]] = 0$.

Proof. This is an easy application of the definition above. ■

We note that these properties make the real space of vector fields on M into a Lie algebra, as defined in Section 9.6.

For the local expression of the Lie bracket, let $(x_1, \cdots, x_m)$ be a local system of coordinates in the open subset U of M, and let X be a vector field on M. Then, by (2.2) and the definition of Xf, for each smooth function f on M

$$Xf\big|_U = a_i \frac{\partial f}{\partial x_i},$$

where a_i is the coefficient of $\frac{\partial}{\partial x_i}$ in the local expression for X. If Y is represented by $b_j \frac{\partial}{\partial x_j}$, it follows from the derivation property that the local expression for the Lie bracket $[X, Y]$ is

$$\left(a_i \frac{\partial b_j}{\partial x_i} - b_i \frac{\partial a_j}{\partial x_i} \right) \frac{\partial}{\partial x_j}.$$

Note the special case $\left[\frac{\partial}{\partial x_i}, \frac{\partial}{\partial x_j} \right] = 0$. Indeed this property characterises coordinate fields.

Theorem 2.6.3. *Given vector fields $X_1, \cdots, X_m$ on an open subset U of a manifold M^m, there is a chart $\phi : U \to \mathbb{R}^m$ such that $\phi_*(X_i) = \frac{\partial}{\partial x_i}$, $i = 1, 2, \cdots, m$, if and only if $[X_i, X_j] = 0$ for all i and j.*

Proof. This result is due to Frobenius. See Warner (1983) for details. ■

2.7. Integral Curves & 1-parameter Groups of Diffeomorphisms.

We conclude this chapter with a more detailed look at vector fields. Firstly we see how vector fields may be explored and characterised by their 'integral curves' and then how, under reasonable hypotheses, these integral curves fit together to give rise to families of diffeomorphisms of the manifold.

Definition 2.7.1. Given a vector field X on the manifold M, a curve $\gamma : (a, b) \to M$ is an *integral curve of* X if $\gamma'(t) \equiv \gamma_*(t)\cdot 1$ is equal to $X(\gamma(t))$ for all $t \in (a, b)$.

Remark 2.7.2. Integral curves may also be referred to as 'flow lines'. When an integral curve is drawn on a plane domain for a vector field on the plane, the tangent vector at each point indicates the direction of the vector field at that point. However, unlike the 'arrow' representation mentioned above, there is no visual representation of the magnitude, since that requires knowledge of the parameterisation of the curve.

As these representations imply, it is always possible to find integral curves of a vector field to fill at least some neighbourhood of a given point of the manifold.

Proposition 2.7.3. *Given a vector field X on a manifold M and a point $x \in M$, there are an open neighbourhood U of x in M, an interval $(-\epsilon, \epsilon)$ of the real line and a unique map $F : U \times (-\epsilon, \epsilon) \to M$ such that, for each $y \in U$, the map $\gamma_y : (-\epsilon, \epsilon) \to M$ given by $\gamma_y(t) = F(y, t)$ is an integral curve of X such that $\gamma_y(0) = y$.*

Proof. Working on a coordinate neighbourhood U of M containing x, with a chart ϕ, the requirement is equivalent to finding $G : \phi(U) \times (-\epsilon, \epsilon) \to \mathbb{R}^m$ such that, for each $z \in \phi(U)$, the map $\gamma_z : (-\epsilon, \epsilon) \to \mathbb{R}^m$ given by $\gamma_z(t) = G(z, t)$ is an integral curve of $\phi_*(X)$ such that $\gamma_z(0) = z$. Writing $\phi_*(X)$ as $\xi_j \partial_j$ in terms of the basis vector fields obtained in the previous section, this is equivalent to solving the m first order linear differential equations:

$$\frac{\partial G_j}{\partial t}(z,t) = \xi_j(z), \ j = 1, \cdots, m,$$

subject to the initial conditions $G_j(z, 0) = z_j$. The existence of a unique solution to such equations on a, possibly smaller, product neighbourhood of any given point is a standard result from the theory of differential equations. See Chapter 9 for more details. ■

The map F in the proposition may also be referred to as a *local flow* of the vector field. For this result it is, of course, not necessary to work with an interval that is symmetric about the origin, nor even with one that includes the origin. The uniqueness clause remains valid for any choice of interval and is very important. It implies in particular that integral curves are unique in the following sense. If $\gamma : (a, b) \to M$ and $\delta : (c, d) \to M$ are integral curves of X such that $a < c < b < d$ and if $\gamma(t_0) = \delta(t_0)$ for some t_0 such that $c < t_0 < b$, then $\gamma(t) = \delta(t)$ for all t such that $c < t < b$ and the 'union'

$$\epsilon(t) = \begin{cases} \gamma(t) & \text{for } a < t < b, \\ \delta(t) & \text{for } c < t < d \end{cases}$$

is also an integral curve for X. Moreover, if γ_x is an integral curve starting at $x = \gamma_x(0)$, $y = \gamma_x(t)$ and τ_t is the translation $s \mapsto s + t$, then $\gamma_y = \gamma_x \circ \tau_t$ is an integral curve, hence *the* integral curve, starting at $y = \gamma_y(0)$. If we now change our point of view and, instead of fixing x and letting t vary, we fix t and let x vary, we obtain a mapping of U into M:

$$\rho_t : U \longrightarrow M; \quad x \mapsto \gamma_x(t).$$

Then, in particular, ρ_0 is the identity and the above identification of integral curves translates to give $\rho_s(\rho_t(x)) = \rho_{s+t}(x)$, at least locally, wherever the composite is defined. In the special case that the flow in Proposition 2.7.3 is global, that is, it is defined on $M \times \mathbb{R}$, we have the following.

Definition 2.7.4. A 1-*parameter group of diffeomorphisms of a manifold* M is a (smooth) map $\rho : M \times \mathbb{R} \to M$ such that, for all $x \in M$ and $t \in \mathbb{R}$, $\rho(x, s+t) = \rho(\rho(x,t), s)$.

Defining $\rho_t(x) = \rho(x, t)$, we note that this means that $\rho_s \circ \rho_t = \rho_{s+t}$ and this immediately implies that ρ_0 is the identity and then that ${\rho_t}^{-1} = \rho_{-t}$. Thus we do indeed have a group of diffeomorphisms of M, isomorphic with the additive group of real numbers.

Conversely, given such a 1-parameter group of diffeomorphisms, we may re-interpret it as a family of curves γ_x, defined for each $x \in U$ by $\gamma_x(t) = \rho_t(x)$, and then recover the vector field of which these are the integral curves as $X(x) = \gamma_x'(0)$. For then, if $y = \gamma_x(t)$, the group property becomes $\gamma_y = \gamma_x \circ \tau_t$ and then we have $X(y) = \gamma_y'(0) = \gamma_x'(t)$ as required for γ_x to be an integral curve.

Although a 1-parameter group of diffeomorphisms always determines a vector field in this way, which is said to *generate the group*, vector fields in general only determine a local version of a 1-parameter group. A sufficent condition for a full group is as follows.

Lemma 2.7.5. *A (smooth) vector field on a manifold M that vanishes outside a compact subset K generates a 1-parameter group of diffeomorphisms.*

Proof. The local flow of Proposition 2.7.3 shows that each point of M has a neighbourhood U on which the equation

$$\frac{d\rho_t(x)}{dt} = X(\rho_t(x)), \qquad \rho_0(x) = x$$

has a unique solution, smooth in both variables, provided $|t| < \epsilon$, for some $\epsilon > 0$. Since finitely many such neighbourhoods U cover K, we may let $\epsilon_0 > 0$

be the least of the corresponding ϵ's. Then, since there is the unique solution $\rho_t(x) = x$ outside K, we obtain a unique solution on M for all $t < \epsilon_0$. Moreover this solution satisfies the group law $\rho_{s+t} = \rho_s \circ \rho_t$ provided all of $|s|$, $|t|$ and $|s+t|$ are less than ϵ_0. Then, picking η such that $0 < \eta < \epsilon_0$, we may write any $t > \epsilon_0$ as $p\eta + \delta$ with $0 \leqslant \delta < \eta$ and define $\rho_t = (\rho_\eta)^p \circ \rho_\delta$ and then $\rho_{-t} = {\rho_t}^{-1}$. This gives us the full 1-parameter group of diffeomorphisms. ∎

Example 2.7.6. Let X be the standard coordinate field $\frac{d}{dt}$ on $(-\pi/2, \pi/2)$ and

$$Y(\tan(y)) = D(\tan)(y) \cdot X(y)$$

define the corresponding vector field on $\mathbb{R}$. Then Y does not generate a 1-parameter group of diffeomorphisms of $\mathbb{R}$ since $\frac{d}{dt}$ does not generate one of $(-\pi/2, \pi/2)$. Thus completeness is insufficient to replace compactness as a hypothesis for the lemma.

2.8. Exercises for Chapter 2.

2.1. Show that the derivations of $\mathcal{C}^\infty(M)$ at p form a vector space $\mathcal{D}_p(M)$ isomorphic with $\tau_p(M)$. (See Example 2.5.1 for the definition of a derivation at p.)

2.2. The manifold M^m is said to have a *trivial tangent bundle* if there is a diffeomorphism $\theta : \tau(M) \to M \times \mathbb{R}^m$ such that $\pi_1 \circ \theta = \pi$ and, for all p in M, the restriction $\theta|_{\tau_p(M)}$ is linear. Show that

(*i*) $\mathbf{S}^1$ has a trivial tangent bundle.

(*ii*) M^m has trivial tangent bundle if and only if there are m vector fields $X_1, \cdots, X_m$ on M such that at each $p \in M$ the set of vectors $\{X_1(p), \cdots, X_m(p)\}$ is a basis of $\tau_p(M)$.

2.3. Let A be a skew-symmetric $m \times m$ matrix. Show that $\gamma(t) = \exp(tA) = \sum_{n=0}^{\infty} t^n A^n/n!$ defines a smooth curve in $SO(m)$ and find $\gamma'(0) = [\gamma]_0$ the tangent vector defined by γ at $\gamma(0)$. What are $\tau_I(SO(m))$ and $\tau_g(SO(m))$ where I is the identity and g a general element?

2.4. Extend Whitney's Theorem 2.4.1 to paracompact manifolds.

CHAPTER 3

FIBRE BUNDLES

In this chapter we put the construction of the tangent bundle $\tau(M)$ in a more general setting, that of topological spaces which are locally but not globally products. A simple example is provided by the Möbius strip (identify the two ends of a strip of paper after one twist); many arise in differential geometry, and it is important to give a systematic discussion. We start with local triviality, the local product structure, whose essential properties we encapsulate in the concept of a coordinate bundle. This will enable us to give a more complete and coherent account of the range of vector bundles associated with the tangent bundle of a manifold than was possible in Chapter 2. This in turn will enable us to describe a broader range of manifolds than we could before.

In this chapter, particularly in Section 3.4, we give less detail than elsewhere in the book. Omitting it will not prejudice understanding of the ensuing chapters.

3.1. Coordinate Bundles.

For this, and most of the next, section it will be more convenient to work in the category of topological spaces and continuous maps. Accordingly we relax our global hypothesis that our spaces be smooth manifolds and that our maps be differentiable.

Definition 3.1.1. A *locally trivial fibration* is a quadruple (p, E, B, F) in which E, B and F are topological spaces and $p : E \to B$ is a continuous surjective map such that, for each point $x \in B$, there exists a neighbourhood U of x and a homeomorphism

$$\varphi_U : U \times F \cong p^{-1}(U),$$

which is well-defined on fibres.

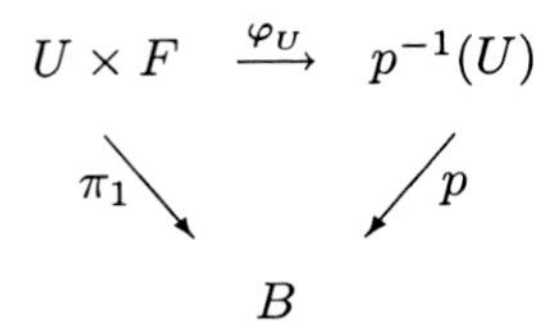

Diagram 3.1.

By 'fibre' we mean any of the sets $p^{-1}(x)$ or $\{x\} \times F$ and the last condition means that φ_U maps $\{x\} \times F$ homeomorphically onto $p^{-1}(x)$ or, put another way, that the triangle in Diagram 3.1 is commutative.

We refer to F itself as the *fibre*, E as the *total space*, B as the *base space* and p as the *projection map* of the fibration. If $U = B$ we say that the homeomorphism φ_U is a *trivialisation* of the fibration in that it provides a fibre-preserving identification of E with the product space $B \times F$.

It follows from the definition that, if $x \in U \cap V$, the composition $\varphi_V^{-1} \circ \varphi_U :$ $\{x\} \times F \to \{x\} \times F$ is a homeomorphism of the fibre F. In order to make the definition more precise and to exercise some control over the homeomorphisms of F which are allowed, it is usual to require the neighbourhood U of x to belong to some family of open subsets $\{U_i \mid i \in I\}$ which cover the base B. The maps $\varphi_i = \varphi_{U_i}$ are then called *coordinate functions* and, with $x \in U_i \cap U_j$ and $\varphi_{i,x} = \varphi_i\big|_{\{x\} \times F}$, we obtain the *coordinate transformations*

$$g_{ji} : U_i \cap U_j \longrightarrow \text{Homeo}(F); \quad x \mapsto \varphi_{j,x}^{-1} \circ \varphi_{i,x}.$$

As x varies we would like this map to be continuous with respect to a suitable topology on Homeo(F) regarded as a subspace of the function space of mappings of the fibre F to itself. It turns out that one suitable topology is the so-called *compact-open*, or C-O, *topology*. If U and K are open and compact subsets of F respectively, write $W(K, U) = \{h \mid h : F \to F \text{ such that } h(K) \subset U\}$. Then, as K and U vary, the subsets W form a subbase of open sets in the function space, that is, an open subset is a union of finite intersections of such special subsets. The usefulness of this definition is shown by the following properties, the proof of which may be found in standard references such as Kelley (1977).

(*i*) If F is Hausdorff, so is Homeo(F).

(*ii*) The continuity of φ_i and φ_j implies that of g_{ij}.

(*iii*) If F is regular and locally compact, then the evaluation map

$$\text{Homeo}(F) \times F \to F;\ (g, x) \mapsto g(x)$$

is continuous.

(*iv*) If F is Hausdorff and compact, then inversion of homeomorphisms is also continuous.

Thus, under restrictions on the fibre F, which are realised in many important examples, Homeo(F) becomes a topological group acting on the space F, where we recall that for a group H to act on a set X requires that, for all h_1, $h_2 \in H$ and $x \in X$, $h_1(h_2(x)) = (h_1h_2)(x)$ and $1_H(x) = x$. However even for the nicest of spaces, such as $\mathbf{S}^n$, the group of all homeomorphisms or even just of all diffeomorphisms is too large, so we restrict to some more tractable subgroup G. An alternative form of words is to represent the topological group G in Homeo(F), that is, to specify a homomorphism of G into Homeo(F), and to say that G acts *effectively* on F if the kernel of this representation is trivial. Having started with no more than a local product structure, we can now formally define a coordinate bundle.

Definition 3.1.2. A *coordinate bundle* comprises the data

$$(p, E, B, F, G, \mathcal{U}, \{\phi_i\}),$$

where p is a continuous map from the total space E to the base space B; $\mathcal{U} = \{U_i \mid i \in I\}$ denotes an open covering of B; G is a topological group acting effectively on the fibre F; and the $\varphi_i : U_i \times F \to p^{-1}(U_i)$ are coordinate functions for the members of $\mathcal{U}$. As for a locally trivial fibration we still require that these coordinate functions be fibre-preserving, but now we also require that the corresponding coordinate transformations define continuous maps $g_{ij} : U_i \cap U_j \to G$, for each pair of members of $\mathcal{U}$.

The main difference with Definition 3.1.1 is that we now require the coordinate transformations to belong to some restricted group of homeomorphisms: if $\rho : G \to \text{Homeo}(F)$ denotes the representation of G as homeomorphisms of the fibre, then that requirement is that $\varphi_{j,x}^{-1} \circ \varphi_{i,x} = \rho(g_{ij})$, for all x in $U_i \cap U_j$. Since G acts effectively, this is equivalent to asking that the coordinate transformations lie in $\rho(G)$. The group G, or $\rho(G)$, is referred to as the *structure group* of the bundle.

As with the differentiable manifolds in Chapter 1, we can remove the dependence on a particular covering $\mathcal{U}$ either by requiring this (together with the coordinate functions) to be maximal, or by working with compatible families $\mathcal{U}$ and $\mathcal{U}'$. Here compatible means that the 'mixed' coordinate transformation $\overline{g}_{ij} : U_j \cap U_i' \to G$ is still continuous and still takes values in the chosen structural group G. We refer to the resulting object as a *fibre bundle*, but in practice we usually work with a representative coordinate bundle adapted to the problem under consideration. It is also common to abbreviate the notation for a fibre bundle to $p : E \to B$, since this determines the fibre up to homeomorphism and the group is often implicit in the context.

Examples 3.1.3. Here are some typical fibre bundles.

1) The product $B \times F$ with trivial structural group and a single coordinate function given by the identity.

2) The (open) Möbius strip with $B = \mathbf{S}^1$, $F = \mathbb{R}$ and $G = \mathbb{Z}/2$. Here we identify $\mathbf{S}^1$ with $\{z \in \mathbb{C} \mid |z| = 1\}$, and describe two domain charts for coordinate functions, $U_1 = \mathbf{S}^1 \setminus \{1\}$ and $U_2 = \mathbf{S}^1 \setminus \{-1\}$. The intersection $U_1 \cap U_2$ has two components and the coordinate transformation g_{21} is given by $g_{21}(x) = \pm 1$ depending on the component in which x lies. There is a similar description of the Klein bottle with fibre $\mathbb{R}$ replaced by fibre $\mathbf{S}^1$. The determination of a coordinate bundle from its coordinate transformations is described below.

3) (Hopf fibration of $\mathbf{S}^3$.) Write $\mathbf{S}^3 = \{(z_1, z_2) \in \mathbb{C}^2 \mid z_1\overline{z}_1 + z_2\overline{z}_2 = 1\}$, identify $\mathbf{S}^2$ with $\mathbb{C}\mathrm{P}^1$ and label a point in projective space by its homogeneous coordinates $[z_1, z_2]$, that is, $[z_1, z_2] = \lambda[z_1, z_2]$ for all $\lambda \neq 0$ with $[0, 0]$ excluded. The map $(z_1, z_2) \mapsto [z_1, z_2]$ is the projection of a non-trivial fibre bundle with fibre and group $\mathbf{S}^1$. There are similar constructions for $\mathbf{S}^7$ (quaternions) and for $\mathbf{S}^{15}$ (octonians or Cayley numbers).

4) Let H be a closed subgroup in the locally compact finite-dimensional topological group G. Then G has the structure of a fibre bundle over the quotient space G/H consisting of the left cosets gH. The fibre is H with group H acting by left translations. The main step in the proof is to find a local cross-section near the point $x_0 = H$ in G/H. This is a continuous function $s : V \to G$ with $p \circ s(x) = x$ for all points x in some neighbourhood V of x_0.

5) The tangent bundle $\tau(M)$ for a smooth manifold and its associated bundles. These will be described at length below.

Definition 3.1.4. Let $p : E \to B$ and $p' : E' \to B'$ be coordinate bundles having the same fibre F and group G. A *bundle map* is a commutative diagram as in Diagram 3.2 such that the induced map on each fibre $f_x : F_x \to F_{fx}$ is a homeomorphism and the family of transformations $g_{kj}(x) = (\varphi'_{k,fx})^{-1} \circ f_x \circ \varphi_{j,x}$ determines a continuous map $U_j \cap \left(f^{-1}(U'_k)\right) \to G$.

$$\begin{array}{ccc} E & \xrightarrow{\tilde{f}} & E' \\ \downarrow{\scriptstyle p} & & \downarrow{\scriptstyle p'} \\ B & \xrightarrow{f} & B' \end{array}$$

Diagram 3.2.

Remark 3.1.5. It is not hard to see that, given $f : B \to B'$, we can always lift f to a bundle map $\tilde{f} : E \to E'$ between total spaces. Furthermore, if f is a homeomorphism, then explicit construction of an inverse shows that $\tilde{f}$ is a homeomorphism also.

Here are two particularly important special cases.

1) If $p : E \to B$ is a fibre bundle over B and $B_0 \subseteq B$ is a subspace of the base space, then $p : p^{-1}(B_0) \to B_0$ is a fibre bundle, the *restriction of* E, and the natural inclusion is a bundle map.
2) Given a fibre bundle $p' : E' \to B'$ and a map $f : B \to B'$, as in Diagram 3.2, the total space of the *induced bundle* $f^*(E') = E$ over B consists of all pairs $(x, e) \in B \times E'$ with $f(x) = p'(e)$ in B'. As a topological space $f^*(E')$ is a subspace of the product; its projecton map p is the projection onto the first factor (check that this is surjective). If E' is trivial over the open subset V of B', then $f^*(E')$ is trivial over $U = f^{-1}(V)$, and the fibre $(f^*(E'))_x$ is mapped homeomorphically onto $E'_{f(x)}$. The coordinate transformations are obtained by composing those for E' with the map f restricted to a suitable open subset of B, and $\tilde{f} : (x, e) \mapsto e$ satisfies the conditions for a bundle map.

The induced bundle construction has an important universal property. Thus, if $g : E'' \to E'$ is a bundle map covering $f : B \to B'$, we can factorise the pair (g, f) as shown in Diagram 3.3: on the upper line $e'' \mapsto (p''(e''), g(e'')) \mapsto g(e'')$, and on the lower $p''(e'') \mapsto p''(e'') \mapsto f \circ p''(e'') = p' \circ g(e'')$ by the condition for a bundle map. So long as some attention is paid to group actions on fibres this construction can be generalised, as we shall soon see for vector bundles.

$$\begin{array}{ccccc} E'' & \xrightarrow{(p'',g)} & f^*(E') & \xrightarrow{\pi_2} & E' \\ \downarrow{p''} & & \downarrow{p} & & \downarrow{p'} \\ B & \xrightarrow{\text{id}} & B & \xrightarrow{f} & B' \end{array}$$

Diagram 3.3.

We note finally that bundle maps have the two functorial properties, namely that the identity is a bundle map and that the composition of two bundle maps is again a bundle map.

Definition 3.1.6. A continuous *section* $s : B \to E$ of the coordinate bundle $p : E \to B$ is a map such that $p \circ s = \text{id} : B \to B$. This generalises the definition of vector field and 1-form in Chapter 2.

We observe that, from their definition, the coordinate transformations determined by coordinate bundle satisfy the conditions:

$$g_{kj}(x)g_{ji}(x) = g_{ki}(x) \qquad \text{for all } x \in U_i \cap U_j \cap U_k. \tag{3.1}$$

From these, or directly from the definition, we also see that

$$g_{ii}(x) = \text{id}_G \qquad \text{(put } i = j = k \text{ in (3.1));}$$
$$g_{jk}(x) = (g_{kj}(x))^{-1} \qquad \text{(put } i = k \text{ in (3.1)).}$$

Conversely, given a fibre on which G acts, such functions $g_{ij} : U_i \cap U_j \to G$ satisfying (3.1) determine a coordinate bundle of which they are the coordinate transformations. It is important to understand how to construct the total space of this, essentially unique, coordinate bundle or, put simply, how to glue together products of the form $U_i \times F$ using the transformation functions g_{ij}.

Definition 3.1.7. By a *system of coordinate transformations on X with values in G* is meant a triple $(B, \mathcal{U}, g_{ij})$, where $\mathcal{U} = \{U_i \mid i \in I\}$ is an open cover of the topological space B and $g_{ij} : U_i \cap U_j \to G$ are continuous functions satisfying (3.1).

We have already implicitly constructed the total space in the case of the tangent bundle in Chapter 2. There the g_{ij} are the derivatives of the coordinate transformations for the manifold, the group is $GL(n, \mathbb{R})$ and the fibre $\mathbb{R}^n$. In general we have

Theorem 3.1.8. *If G acts on the topological space F and $(B, \mathcal{U}, g_{ij})$ is a system of coordinate transformations on the space B with values in G, then there exists a coordinate bundle $p : E \to B$, with fibre F, structural group G and corresponding coordinate transformations g_{ij}.*

Proof. Give the indexing set I of the open cover $\{U_i \mid i \in I\}$ the discrete topology. Let $\overline{E}$ be the subspace of the product $B \times F \times I$ comprising all points (x, y, j) such that $x \in U_j$, that is, $\overline{E}$ is the disjoint union

$$\overline{E} = \bigsqcup_j (U_j \times F \times \{j\}).$$

Define an equivalence relation in this 'pre-total space' $\overline{E}$ by

$$(x, y, j) \sim (x', y', k) \Longleftrightarrow x = x' \text{ and } g_{kj}(x)y = y'.$$

Then the total space of our bundle E equals $\overline{E}/\sim$. Write $q : \overline{E} \to E$ for the quotient map, and give E the quotient topology. Define the bundle projection map by $p(x, y, j) = x$, noting that it is independent of the representative of the equivalence class; p is continuous by the properties of the quotient topology. The definition of the coordinate functions is also self-evident:

$$\varphi_j(x, y) = q(x, y, j), \qquad \text{for } x \in U_j \text{ and } y \in F.$$

Continuity is again a consequence of the quotient properties, bijectivity follows from the conditions (3.1) on the g_{ij}. The hardest part of the proof is to show that φ_j is an open map; this amounts to showing that, if V is open in $U_j \times F$, then $q^{-1} \circ \varphi_j(V)$ meets each component $U_k \times F \times \{k\}$ in an open set. The intersection is contained in $(U_j \cap U_k) \times Y \times \{k\}$, which is certainly open in $\overline{E}$ and on which q can be twisted by the coordinate transformation g_{jk} to bring the question of openness back to the index j. Here the answer is obvious. Inspection of the definitions shows that $g_{ji}(x)(y) = \varphi_{j,x}^{-1} \circ \varphi_{i,x}(y)$, concluding the proof. ∎

A slightly weaker definition of equivalence than that used to define a fibre bundle in terms of its representing coordinate bundles is to say that E and E' are equivalent if they are mapped one to the other by a fibre preserving homeomorphism covering the identity on B.

Notation. The bundles E and E' are said to be *associated* if they have the same coordinate transformations $\{g_{ji}\}$. Thus they differ either in that G is represented in the homeomorphism groups of two distinct fibres F and F', or that we are given two distinct actions of the same structural group on the common fibre F. For example, the tangent bundle and cotangent bundle of a smooth manifold that we met in Chapter 2 are associated bundles. If each transformation g_{ji} takes values in the subgroup H of G, the structural group of E is said to admit reductions to H, or to be *reducible* to H. In particular, E is trivialisable if and only if its structural group is reducible to the identity $\{1_G\}$. In the special case that $G = F$ with G acting on itself by left multiplication the bundle is said to be *principal.*

3.2. Vector Bundles.

In differential geometry the most important single fibre bundle is the tangent bundle of a smooth manifold. It is an example of a *vector bundle* in which we take the fibre F to be a finite-dimensional vector space over the real or complex numbers and require the group G to act linearly. For vector bundles, besides being able to describe restriction on the base space B, we can also define it on the fibre, $F \cong \mathbb{R}^n$ (say).

Definition 3.2.1. With the notation already introduced let $E' \subseteq E$ be a subspace such that the base B of E has an open covering $\mathcal{U}$ for each member U of which there is a coordinate function φ_U for which $\varphi_U^{-1}(p^{-1}(U) \cap E') = U \times \mathbb{R}^k \subseteq U \times \mathbb{R}^n$. Then $p : E' \to B$ is itself a vector bundle over B, called a *k-dimensional subbundle of E.*

Example 3.2.2. If $f : E \to E'$ is a fibre preserving map over the identity map of B and if f is linear and of rank k on each fibre, then $\ker(f) = \bigcup_{x \in B} \ker(f_x)$ and $\operatorname{im}(f) = \bigcup_{x \in B} \operatorname{im}(f_x)$ are subbundles of E and E' respectively.

This is intuitively obvious, although the proof is not entirely trivial, see for example Bröcker and Jänich (1982), Lemma 3.5.

Turning to some of the other definitions given for a general fibre bundle, we see that every vector bundle admits at least one section, the so-called *zero-section,* which maps $x \in B$ to the zero vector in the fibre $p^{-1}(x)$. Widening the definition of bundle map, in which we assumed that F and G were constant, it is useful to consider *linear bundle maps* for which we require only that the fibre $p^{-1}(x)$ be mapped linearly into the fibre $p'^{-1}(f(x))$. If g is a linear bundle map covering $f : B \to B'$, then the universal property of the induced bundle still holds; namely g factors through a map into the total space of the induced vector bundle $f^*(E')$, with fibre dimension equal to that of E'.

Assuming that we have made some choice of basis in the fibre F, the coordinate transformations g_{ji} of an n-dimensional vector bundle take their values in the relevant general linear group. If the base space B is a smooth manifold, then we can require g_{ji} to be smooth and give the total space E the structure of a smooth manifold of dimension $\dim(B) + \dim(F)$. This applies in particular to the tangent bundle, of dimension $2n$, and to its associated principal bundle with fibre $GL(n, \mathbb{R})$ and so of dimension $n + n^2$.

In fact Theorem 3.1.8 allows us to give a more direct definition of $\tau(M)$, starting with $F = \mathbb{R}^n$, $G = GL(n, \mathbb{R})$ with its natural action and $\mathcal{U}$ some atlas of coordinate charts for M. The coordinate transformations are then given by the Jacobian matrices of coordinate change in the overlapping neighbourhoods U_i and U_j. This approach does not require an explicit definition of tangent vectors, such as the equivalence classes of curves in small open subsets of M that we used in Chapter 2. It is very close to the alternative definition of $\tau(M)$ given in Section 2.5 of Bröcker and Jänich (1982).

We have already seen in Chapter 2 how, if $f : M \to N$ is a smooth map, the set of induced maps of tangent spaces combine to give a linear map of bundles

$$f_* : \tau(M) \longrightarrow \tau(N).$$

Since smoothness is a local property the smoothness of f implies that of f_* and, by the universal property of induced bundles, there exists a unique map $\tau(M) \to f^*(\tau(N))$ making the Diagram 3.4 commute.

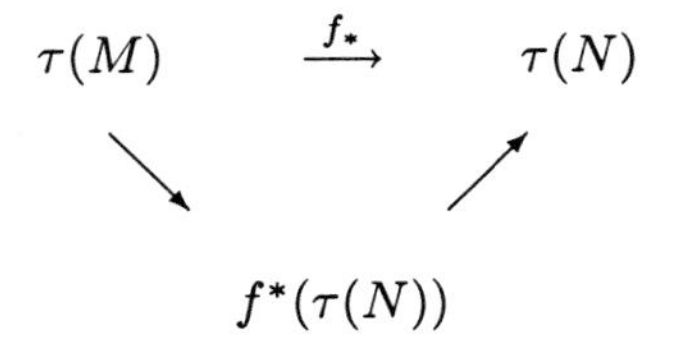

Diagram 3.4.

Now consider the problem of the trivialisation of a vector bundle. It is clear that a necessary condition for triviality is the existence of a section which is nowhere zero. A simple generalisation gives a sufficient condition in terms of a basis of sections, that is, a family of sections $\{s_1, \cdots, s_n\}$ of the $\mathbb{R}^n$-bundle E with the property that for each $x \in B$ the images $s_1(x), \cdots, s_n(x)$ form a basis for the vector space $p^{-1}(x)$.

Theorem 3.2.3. *Let $p : E \to B$ be a vector bundle of fibre dimension n. Then E is trivialisable if and only if E admits a basis of sections. Furthermore in this case every section of E is uniquely expressible as $s = \alpha_j s_j$ where α_j is a continuous real-valued function on B.*

Proof. It is clear that $B \times \mathbb{R}^n$ admits such a basis of sections. Conversely the existence of such a family allows us to construct an isomorphism between E and $B \times \mathbb{R}^n$ by mapping the basis $\{s_1(x), \cdots, s_n(x)\}$ of $p^{-1}(x)$ to a suitable basis of $\mathbb{R}^n$. The second part is clear from the definition of a basis, the continuity of the functions α_j following from that of B. ■

Linear Algebra for Vector Bundles. It is intuitively clear that the usual operations which one carries out on vector spaces and homomorphisms between them extend to vector bundles, provided that the operations are performed fibrewise. For example, and we shall give the details below, the Whitney sum $E \oplus E'$ of two bundles is obtained, at least as a set, by forming the direct sum $E_x \oplus E'_x$ over each point x of the base space.

Definition 3.2.4. Let E and E' be vector bundles over X. If $\{g_{ij}\}$ and $\{h_{\alpha\beta}\}$ are the coordinate transformations for E and E' respectively, then

$$k_{ij\alpha\beta} : U_i \cap U_j \cap V_\alpha \cap V_\beta \to GL(n, \mathbb{R}) \oplus GL(m, \mathbb{R}); x \mapsto g_{ij}(x) \oplus h_{\alpha\beta}(x)$$

defines a system of coordinate transformations for a vector bundle over X, with fibre $\mathbb{R}^n \oplus \mathbb{R}^m$ and trivialising cover $\{W_{i\alpha} = U_i \cap V_\alpha \,\big|\, (i, \alpha) \in I_E \times I_F\}$. This bundle is called the Whitney sum of E and E', written $E \oplus E'$.

It is clear that, if $f_1 : E_1 \to E'_1$ and $f_2 : E_2 \to E'_2$ are linear bundle maps, then we can construct $f_1 \oplus f_2 : E_1 \oplus E_2 \to E'_1 \oplus E'_2$ by taking the obvious map on each fibre. Furthermore differentiability is inherited by sums and maps between them. In the same way we obtain the further examples:

1) the tensor product $E \otimes E'$;
2) the quotient bundle E/E', where E' is a subbundle of E;
3) the dual bundle E^*;
4) the exterior power bundles $\wedge^k E$ and $\wedge^k E^*$, of which more in Chapter 4 below.

Warning example. The Whitney sum of non-trivial bundles may be trivial. Let E be the Möbius band regarded as an $\mathbb{R}$-bundle over $\mathbf{S}^1$. If we choose $U_1 = \{e^{i\theta} | -\pi < \theta < \pi\}$ and $U_2 = \{e^{i\theta} | 0 < \theta < 2\pi\}$ to cover $\mathbf{S}^1$, then $U_1 \cap U_2 = A \cup B$, where $A = \{e^{i\theta} | 0 < \theta < \pi\}$ and $B = \{e^{i\theta} | -\pi < \theta < 0\} \equiv \{e^{i\theta} | \pi < \theta < 2\pi\}$. With these choices, E is determined by the transition function g_{12} where

$$g_{12}(e^{i\theta}) = \begin{cases} 1 & \text{if } 0 < \theta < \pi, \\ -1 & \text{if } \pi < \theta < 2\pi. \end{cases}$$

Hence the Whitney sum $E \oplus E$ is determined by the transition function G_{12} defined by

$$G_{12}(e^{i\theta}) = \begin{cases} I_2 & \text{if } 0 < \theta < \pi, \\ -I_2 & \text{if } \pi < \theta < 2\pi. \end{cases}$$

Now the map $s : e^{i\theta} \mapsto (\cos\theta/2, \sin\theta/2)$ is not well-defined unless we restrict attention to, say, U_1 and U_2. Calling the resulting maps s_1 and s_2, we see that $s_1\big|_A = s_2\big|_A$ while $s_1\big|_B = -s_2\big|_B$. Thus in both cases $s_1 = G_{12}s_2$ and so s is well-defined as a cross-section of $E \oplus E$. Similarly $t : e^{i\theta} \mapsto (-\sin\theta/2, \cos\theta/2)$ determines a never zero cross-section of $E \oplus E$, which is everywhere linearly independent of s, so that together they form a real basis of sections as required by Theorem 3.2.3 to trivialise $E \oplus E$.

Definition 3.2.5. The $\mathbb{R}^n$-bundle over X is *orientable* if it is possible to choose a bundle atlas $\{U_i \mid i \in I\}$ such that whenever $U_i \cap U_j \neq \emptyset$ the homeomorphism $\varphi_j^{-1} \circ \varphi_i$ restricts to a linear map on fibres having an everywhere positive determinent.

The definition just given is equivalent to requiring that the structural group of the given vector bundle be reducible from $GL(n, \mathbb{R})$ to $SL(n, \mathbb{R})$. Any such reduction will be called an *orientation* on the bundle. Note that there exist bundles which cannot be oriented, for example the Möbius band.

Definition 3.2.6. The m-dimensional $\mathcal{C}^\infty$-manifold M^m is said to be orientable if the $\mathbb{R}^m$-bundle $\tau(M)$ over M is orientable.

This amounts to being able to choose an atlas A for M such that the Jacobian matrix $\left(\frac{\partial y_j}{\partial x_i}\right)$ belongs to $SL(n, \mathbb{R})$ for all points $x \in U_i \cap U_j$ and all pairs (i, j). A possibly more elegant equivalent definition of the orientation of manifolds will be given in Chapter 4.

Example 3.2.7. The torus T^2 is orientable, but the Klein bottle is not. Real projective n-space $\mathbb{RP}^n$ is orientable if and only if n is odd. The reader should already be able to establish that T^2 is orientable. Tools for proving the other results will appear in later chapters.

3.3. Riemannian Metrics.

We now turn to the existence of Riemannian metrics. If E_x is a real vector space, then one can consider bilinear forms $E_x \times E_x \to \mathbb{R}$ as elements of the dual space $(E_x \otimes E_x)^*$. If $E = \bigcup_{x \in B} E_x$ is a vector bundle over B then, as for the Whitney sum, we can construct the bundle $(E \otimes E)^*$.

Definition 3.3.1. A *Riemannian metric* for E is a continuous section $s : B \to (E \otimes E)^*$ with the property that for each $x \in B$ the corresponding bilinear form $E_x \times E_x \to \mathbb{R}$ is symmetric and positive definite. The metric is of class $\mathcal{C}^k$, $0 \leqslant k \leqslant \infty$, if B and E are $\mathcal{C}^k$-manifolds and s a $\mathcal{C}^k$-section. We say that the choice of a Riemannian metric g on $\tau(M)$ imposes a *Riemannian structure* on the manifold M^m and refer to the pair (M, g) as a *Riemannian manifold.*

Remark 3.3.2. If one weakens positive definite to non-singular and indefinite, then one obtains what is called a pseudo-Riemannian metric. These arise naturally in general relativity.

Lemma 3.3.3. *If $E_1 \subseteq E$ is a sub-vector bundle of E and E admits a Riemannian metric, then*

$$E_1^{\perp} = \bigcup_{x \in B} E_{1x}^{\perp}$$

is also a subvector bundle, where in each fibre $E_{1x}^{\perp}$ denotes the orthogonal complement of E_{1x} with respect to the inner product $s(x)$.

Proof. If (f, U) is a local chart for E, which represents $E_1\big|_U$ as $U \times (\mathbb{R}^k \times \{0\}) \subset U \times \mathbb{R}^n$, and $v_1, \cdots, v_n$ are sections of $E|_U$ corresponding to the standard basis vectors of $\mathbb{R}^n$, then one can apply the Gram-Schmidt orthogonalisation procedure to obtain sections $v_1', \cdots, v_n'$, such that $v_1', \cdots, v_k'$ still span $E_1|_U$, as did $v_1, \cdots, v_k$, and the remaining sections span $E_1^{\perp}|_U$. In this way

$$f' : E|_U \longrightarrow U \times \mathbb{R}^n; \quad \lambda_i v_i'(x) \mapsto (x, \lambda_1, \cdots, \lambda_n)$$

defines bundle charts both for $E_1\big|_U$ and $E_1^{\perp}\big|_U$. ■

Corollary 3.3.4. *If E admits a Riemannian metric, then its structural group admits reductions to the subgroup $O(n) \subseteq GL(n, \mathbb{R})$ or to $SO(n) \subseteq SL(n, \mathbb{R})$ in the orientable case.*

Proof. This follows immediately from the proof of the lemma. ∎

Lemma 3.3.3 has the following important application – keeping the same notation. Let E/E_1 denote the quotient bundle (intuitively take the quotient on each fibre), then $E_1^{\perp} \cong E/E_1$. Hence the choice of a Riemannian metric allows us to split off a subbundle E_1 from E as a direct summand: compare the theory of representations of a finite group.

Definition 3.3.5. Let $N^n \hookrightarrow M^m$ be a differentiable submanifold. Then the *normal bundle* of the embedding, $\nu(N^n)$, is defined to be the quotient bundle $(\tau(M)|_N)/\tau(N)$ of fibre dimension $m - n$.

Given a Riemannian structure on M, Lemma 3.3.3 implies that $\nu(N^n)$ is isomorphic to $\tau(N)^{\perp}$. See also the remarks preceeding Corollary 7.4.3.

Example 3.3.6. Consider the standard embedding of $\mathbf{S}^2$ in $\mathbb{R}^3$. Since $\tau(\mathbb{R}^3) \cong \mathbb{R}^3 \times \mathbb{R}^3$, the restriction of this tangent bundle to $\mathbf{S}^2$ equals $\mathbf{S}^2 \times \mathbb{R}^3$, isomorphic to $\tau(\mathbf{S}^2) \oplus \nu\mathbf{S}^2$. The second summand is trivial, that is, isomorphic to $\mathbf{S}^2 \times \mathbb{R}$, because by slightly extending the sphere we can construct a nowhere vanishing section and the fibre dimension equals one. However we shall see in Chapter 7 that $\tau(\mathbf{S}^2)$ is not trivialisable: indeed by Corollary 7.6.7, since its Euler characteristic is non-zero, $\mathbf{S}^2$ does not even have one nowhere vanishing section, let alone the two independent ones that would be necessary to trivialise its tangent bundle. This is sometimes called the Hairy Ball Theorem: a hairy ball cannot be combed flat. This example shows that cancellation fails for vector bundles, that is, that the direct sum of E and a trivial bundle can be trivial without E itself being trivial.

Theorem 3.3.7. (Existence of Riemannian Metrics.) *Let M^m be a compact smooth manifold. Then $\tau(M)$ admits Riemannian metrics.*

Proof. Let $\{(U_i, \varphi_i) \mid i \in I\}$ be a finite atlas for M^m, which exists since M^m is compact. Without loss of generality $\tau(M)|_{U_i} \cong U_i \times \mathbb{R}^m$, so that we may put a Riemannian metric on $\tau(M)|_{U_i}$, simply by using the standard inner product on $\mathbb{R}^m$ to define a constant section s_i of $(\tau(M)|_{U_i} \otimes \tau(M)|_{U_i})^*$. Now let $\{\chi_i \mid i \in I\}$ be the smooth partition of unity constructed in Section 1.5 subordinate to the given cover $\{U_i \mid i \in I\}$. If we define $\chi_i(x)s_i(x) = 0$ for

$x \notin \text{supp}(\chi_i)$, then $\chi_i s_i$ becomes a smooth global section of $(\tau(M) \otimes \tau(M))^*$. Write $s = \chi_i s_i$ for the required Riemannian metric. This section is both symmetric and positive definite, because these are *convex* properties: the fact that A and B are symmetric and positive definite implies the same for all matrices lying on the line segment $tA + (1-t)B$, $0 \leqslant t \leqslant 1$, joining them. ■

The result that we have just proved shows both the usefulness and the limitations of partitions of unity. One can always use the functions $\{\chi_i\}$ to glue local sections of a bundle associated to $\tau(M)$ together. However the resulting global section may not have the required properties unless local sections having those properties form a convex set. We shall see later that orientation can be expressed in terms of sections of the exterior power bundle $\wedge^m \tau(M)$ and, although it is clear that local sections with no zeros always exist, the same may not be true globally. The same is true for vector fields.

If (M, g) and (N, h) are Riemannian manifolds and $f : M \longrightarrow N$ a smooth map, then for each point x in M the derivative f_* induces a map

$$f^* : (\tau_{fx}N \times \tau_{fx}N)^* \longrightarrow (\tau_x M \times \tau_x M)^*,$$

where $(f^*(\beta))(v, w) = \beta(f_*(v), f_*(w))$.

Definition 3.3.8. A smooth map $f : M \longrightarrow N$ between Riemannian manifolds (M, g) and (N, h) is *isometric* if, for all $x \in M$, $f^*(h_{fx}) = g_x$. An isometric diffeomorphism is called an *isometry*.

A Riemannian structure g on a manifold M enables one to define the length of curves in M: the curve $\gamma : [a, b] \longrightarrow M$, where $[a, b] \subset \mathbb{R}$, has length $L[\gamma]$ defined by

$$L[\gamma] = \int_a^b \sqrt{g_{\gamma(t)}(\gamma'(t), \gamma'(t))}\, dt.$$

Definition 3.3.9. The curve $\gamma : [a, b] \longrightarrow M$ is called a *geodesic* if $L[\gamma] \leqslant L[\delta]$ for all curves $\delta : [c, d] \longrightarrow M$ such that $\gamma(a) = \delta(c)$ and $\gamma(b) = \delta(d)$, that is, for all curves δ joining the same two points of M.

Examples 3.3.10. (1) If, for each $p \in \mathbb{R}^m$, we give $\tau_p(\mathbb{R}^m)$ the standard inner product, the formula for the lengths of curves under the resulting Riemannian structure is the familiar one and so the geodesics are the straight lines.

(2) More generally, if a group G of isometries acts on the Riemannian manifold (M, g) in such a way that M/G inherits a differential structure (see,

for example, Exercise 1.6) then it also inherits a Riemannian structure such that the quotient map is isometric. As a simple example the group G generated by the translation $t \mapsto t + 2\pi$ of $\mathbb{R}$ with the standard Riemannian structure induces a Riemannian structure on the quotient space, which we may identify with the unit circle $\mathbf{S}^1$ in the complex plane. Then the quotient map is the exponential: $t \mapsto e^{it}$. This maps geodesics of length at most π, but not longer ones, onto geodesics in $\mathbf{S}^1$. The image geodesics have the same length so that, if there is one geodesic of length π joining two points in $\mathbf{S}^1$, then there are always two.

3.4. Applications.

In the final section of this chapter we first prove a theorem which gives a sufficient condition for a smooth surjective map $f : E^m \to M^n$ $(m > n)$ to be a locally trivial fibration. In fact we already know, from the Submersion Theorem, 1.3.2, that this is true locally along each fibre: each fibre F is a submanifold. What we need to show now is that it has a global product neighbourhood $F \times U$. There is a sense in which this is 'dual' to the result proved in Section 1.3 stating that an injective immersion of a compact manifold is an embedding. Indeed the compactness assumption plays a similar rôle in both cases, and can be weakened, provided some restriction is placed on the map f.

Our second aim is to show the rôle of fibrations in the structure of higher, mainly three, dimensional manifolds, and to give examples of 'irreducible' manifolds which behave like copies of the torus T^2 in dimension two. Indeed it is an attractive conjecture that, by cutting along suitable copies of $\mathbf{S}^2$ and T^2, any connected, compact, closed oriented 3-manifold can be decomposed into a collection of building blocks, each of which is related to one of the examples we describe.

Theorem 3.4.1. (C. Ehresmann.) *Let $f : E \to M$ be a smooth submersion of a compact smooth manifold E onto M. Then f is a smooth locally trivial fibration, that is, if $x \in M$ and $F = f^{-1}(x)$, then there exists some neighbourhood U of x in M and a diffeomorphism $\varphi : U \times F \to f^{-1}(U)$ such that Diagram 3.5 is commutative:*

Proof. Since the result is local in M, we may, without loss of generality, take $M = \mathbb{R}^m$ with coordinates $x_1, \cdots, x_m$ near $p = \underline{0}$, and basic tangent vector

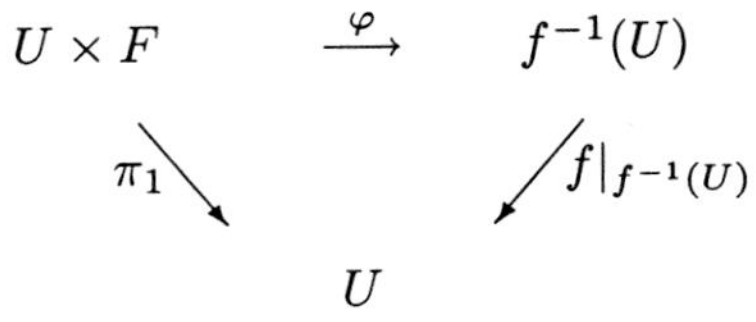

$$\begin{array}{ccc} U \times F & \xrightarrow{\varphi} & f^{-1}(U) \\ & \pi_1 \searrow \quad \swarrow f|_{f^{-1}(U)} & \\ & U & \end{array}$$

Diagram 3.5.

fields $\frac{\partial}{\partial x_j}$, $j = 1, \cdots, m$. Lift these vector fields to $v_1, \cdots, v_m$ on E, such that $f_*(v_j) = \frac{\partial}{\partial x_j}$. We can achieve this by applying Corollary 1.3.7 to see that the submersion f locally has the form $\pi_1 : U_k \times V_k \to U_k$, giving obvious vector fields v_{jk}, $j = 1, \cdots, m$, on $U_k \times V_k$. For each j we may then glue these together by means of a partition of unity, subordinate to the covering of E by the $U_k \times V_k$. Note that, since f_* is linear it will still map the resulting vector fields v_j to $\frac{\partial}{\partial x_j}$ on $\mathbb{R}^m$.

If the fibre in which we are interested is $F = f^{-1}(0)$ then, by means of a bump function on $\mathbb{R}^n$ that takes the value 1 on a closed ball $\bar{B}$ centred on the origin and is supported on $2\bar{B}$, we may assume that the vector fields are supported on the closed and, hence, compact set $f^{-1}(2\bar{B})$ and have the required projections on $f^{-1}(\bar{B})$. Then, by Lemma 2.7.5, each vector field v_j generates a 1-parameter family of diffeomorphisms Φ^j on E. Fixing y in F, the m flow lines from y outwards given by $\Phi^1, \cdots, \Phi^m$ correspond to local coordinate axes in the base manifold M^m throughout $f^{-1}(\bar{B})$. Thus, for $y \in F$ and $\underline{x} = (x_1, \cdots, x_m) \in U = \text{int}(\bar{B}) \subset M$, we may define $\varphi(\underline{x}, y) = \Phi^1_{x_1} \circ \cdots \circ \Phi^m_{x_m}(y)$.

This is indeed the required local trivialisation, that is,

$$\underline{x} = \pi_1(\underline{x}, y) = f \circ \varphi(\underline{x}, y),$$

since $f \circ \Phi^j_{x_j}(z) = f \circ \Phi^j_0(z) + x_j e_j$, where $e_j \in \mathbb{R}^m$ is the jth unit vector. This latter equation holds for $x_j = 0$, and the agreement of the derivatives according to x_j is guaranteed, since v_j lifts $\partial/\partial x_j$. Note that for $f(y) = \underline{x}$ the inverse diffeomorphism is given by reversing the flow, that is, $\varphi^{-1}(y) = (\underline{x}, \Phi^m_{-x_m} \cdot \cdots \cdot \Phi^1_{-x_1}(y))$. ∎

In the proof above the compactness assumption on E was necessary to ensure that the closed set $f^{-1}(\bar{B})$ was compact which, in turn, ensured that each flow line extended as far as it was needed. In general, without alternative hypotheses, the sets U_k over which the local product and, hence, the required vector fields are defined might intersect only in a single point.

In giving our family of representative 3-dimensional manifolds, let us first recall that in dimension two a connected, compact, closed oriented surface is built up from copies of the sphere $\mathbf{S}^2$ and torus $T^2 = \mathbf{S}^1 \times \mathbf{S}^1$ by means of the operation of connected sum. (If you have not seen this before, turn to Exercise 6.5 at the end of Chapter 6.) Then for each such surface F we can identify a simply connected, but not necessarily compact, 'model' Riemannian surface with a group of isometries such that the quotient surface is diffeomorphic with F, which therefore inherits a correponding Riemannian structure.

We distinguish three cases: $\mathbf{S}^2$, the unit sphere in $\mathbb{R}^3$, with the standard Riemannian structure, which is its own model; the plane $\mathbf{E}^2$, again with the standard Riemannian structure, which is the model for the compact surface T^2; and the hyperbolic plane $\mathbf{H}^2$, which is the model for the connected sum of g copies of the torus, with $g \geqslant 2$. In Diagram 3.6 we give the Riemannian metric associated with each of these model spaces in terms of local coordinates. They are also characterised by their 'parallel lines', that is, geodesics which never meet. In the last, the Poincaré model for the hyperbolic plane, straight lines are given by arcs of circles meeting the boundary circle

$$C = \{(x, y) \mid x^2 + y^2 = 1\}$$

orthogonally. This set includes Euclidean lines passing through $(0, 0)$. Taking ℓ to be such a line and $p \notin \ell$, it is easy to find infinitely many lines through p not meeting ℓ inside the disc. Of these only the two that meet ℓ on C are regarded as parallel and the remainder are termed *ultraparallel.*

$\mathbf{S}^2$	$d\theta^2 + \sin^2\theta d\varphi^2$	polar coordinates $(r = 1, \varphi, \theta)$,	no parallel lines
$\mathbf{E}^2$	$dx^2 + dy^2$	(x, y) Cartesian coordinates,	unique line parallel to a given line ℓ through $p \notin \ell$
$\mathbf{H}^2$	$\frac{4(dx^2+dy^2)}{(1-(x^2+y^2))^2}$	coordinates inside the unit disc,	two lines parallel to ℓ through $p \notin \ell$. (See text.)

Diagram 3.6.

Theorem 3.4.2. *Each connected, closed, compact orientable surface admits a Riemannian metric that is locally isometric to one of* $\mathbf{S}^2$, $\mathbf{E}^2$, $\mathbf{H}^2$.

Proof. Let F_g^2 denote an oriented surface of genus g. For $F_0^2 = \mathbf{S}^2$ the conclusion is obvious; the non-orientable surface $\mathbb{R}\mathrm{P}^2$ has the same geometry. $F_1^2 = \mathbf{S}^1 \times \mathbf{S}^1 \cong \mathbf{E}^2/\mathbb{Z} \times \mathbb{Z}$, where the group $\mathbb{Z} \times \mathbb{Z}$ is generated by translations, which are isometries, through unit distance in the x and y directions. Now consider F_2^2.

Let Y be a regular octagon embedded in $\mathbf{H}^2$ with angles equal to $\pi/4$. Define isometries $\alpha, \beta, \gamma, \delta$ of $\mathbf{H}^2$ by the rules $\alpha(AB) = DC$, $\beta(BC) = ED$, $\gamma(EF) = HG$, $\delta(FG) = AH$, and in all cases require that the interior of Y be disjoint from its translates. (See Figure 3.7.) Then, given the description of F_2^2 in the classification of surfaces, if $\Gamma_2 = \langle \alpha, \beta, \gamma, \delta \rangle$, $\mathbf{H}^2/\Gamma_2$ is homeomorphic to F_2^2.

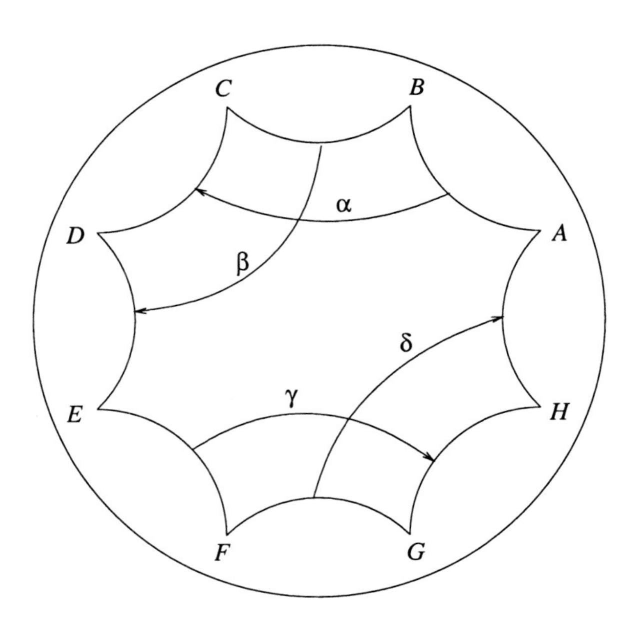

Figure 3.7.

In order to deal with surfaces of higher genus we exhibit $\Gamma_g = \pi_1(F_g^2)$, $g \geqslant 3$, as a subgroup of $\pi_1(F_2^2)$, and then form the orbit space of $\Gamma_g \subset \Gamma_2$. ■

In dimension three we try to generalise this construction, starting with the models $\mathbf{S}^3, \mathbf{E}^3$ and $\mathbf{H}^3$. The next step is to take the products $\mathbf{S}^2 \times \mathbf{E}^1, \mathbf{E}^2 \times \mathbf{E}^1 = \mathbf{E}^3$ and $\mathbf{H}^2 \times \mathbf{E}^1$ with compact models $\mathbf{S}^2 \times \mathbf{S}^1$, $T^3 = \mathbf{S}^1 \times \mathbf{S}^1 \times \mathbf{S}^1$ and $F_g^2 \times \mathbf{S}^1$ ($g \geqslant 2$), respectively. These five examples are augmented by three more, corresponding to taking non-trivial fibrations, either over F_g^2, $g \geqslant 0$, or over $\mathbf{S}^1$. We obtain the table in Diagram 3.8, where a bundle over $\mathbf{S}^1$ is isomorphic to the mapping torus, $E(h)$, of a diffeomorphism $h : F^2 \to F^2$, where $E(h) = F^2 \times [0,1]/(x,0) \sim (h(x),1)$. For example, $E(\mathrm{id}) = F \times \mathbf{S}^1$. In the table we only specify h when it is not the identity.

Dim 2		Dimension 3			
model	*type*		*compact examples*		*model*
$\mathbf{S}^2$	Elliptic	$\mathbf{S}^2$	$\mathbf{S}^1$-bundle over $\mathbf{S}^2$	$\mathbf{S}^1 \times \mathbf{S}^2$ $\mathbf{S}^3$	$\mathbf{E}^1 \times \mathbf{S}^2$ $\mathbf{S}^3$
$\mathbf{E}^2$	Flat or Euclidean	T^2	T^2-bundle over $\mathbf{S}^1$	$\mathbf{S}^1 \times T^2$ $h \in SL(2,\mathbb{Z})$	$\mathbf{E}^1 \times \mathbf{E}^2 = \mathbf{E}^3$ Nil^3 or Sol^3
$\mathbf{H}^2$	Hyperbolic	F^2	F^2-bundles over $\mathbf{S}^1$:	$\mathbf{S}^1 \times F^2$ h is pseudo-Anosov	$\mathbf{E}^1 \times \mathbf{H}^2$ $\mathbf{H}^3$
			$\mathbf{S}^1$-bundles over F^2 ($\hookrightarrow \tau(F^2)$)		$\widetilde{SL}(2,\mathbb{R})$.

Diagram 3.8.

Remarks 3.4.3. 1) Three-dimensional elliptic manifolds are modeled on $\mathbf{S}^3$, the total space of the Hopf fibration described earlier in this chapter.

2) The generalisation of a flat surface is a manifold of Lie type. The diffeomorphism of T^2 is obtained by allowing the matrix h to act on $\mathbb{R}^2$ in the standard way and then taking the induced action on $\mathbb{R}^2/\mathbb{Z}^2$. Assuming that $h \neq$ identity we distinguish between two cases, nilpotent $\left(\begin{pmatrix} a & b \\ c & d \end{pmatrix} = \begin{pmatrix} 1 & b \\ 0 & 1 \end{pmatrix}\right)$ and solvable (arbitrary).

3) The 'pseudo-Anosov' assumption is technical and is needed to give the total space of the fibration a hyperbolic structure. As with surfaces hyperbolic 3-manifolds are generic, and not all examples are related to these fibrations. It has however been asked whether a hyperbolic manifold is finitely covered by an F^2-bundle over $\mathbf{S}^1$. The $\mathbf{S}^1$-bundles over surfaces F^2 are obtained by putting a Riemannian metric on the tangent bundle $\tau(F^2)$, and defining a subbundle by taking as fibre the unit circle in each 2-dimensional tangent space.

The symbols Nil^3 and Sol^3 stand for 3-dimensional nilpotent and solvable Lie groups respectively. Lie groups in general will be introduced in Chapter 8 below; for the moment we note that they provide the simply connected models for two of our examples above. Nil is a non-split central extension of $\mathbb{R}$ by $\mathbb{R}^2$, which can be identified with the matrix group

$$\left\{ \begin{pmatrix} 1 & x & z \\ 0 & 1 & y \\ 0 & 0 & 1 \end{pmatrix} \middle| \, x, y, z \in \mathbb{R} \right\}.$$

The central subgroup $\mathbb{R}$ corresponds to matrices such that $x = y = 0$. Finally Sol is another 3-dimensional Lie group, this time a split extension of $\mathbb{R}^2$ by $\mathbb{R}$, where if t generates the quotient, the action of t on $\mathbb{R}^2$ is given by

$$(x, y) \mapsto (e^t x, e^{-t} y).$$

Note that t defines a linear isomorphism of $\mathbb{R}^2$ to itself with determinant 1 and distinct real eigenvalues.

How complete do we believe this list of 3-dimensional model spaces to be? In general we say that the manifold M^m is *geometric* if it is locally isometric to the simply connected Riemannian manifold (X^m, g) with a group $\mathfrak{I}(X)$ of isometries acting *transitively.* This means that, given two points x, $y \in X$, there is an isometry $f \in \mathfrak{I}(X)$ such that $f(x) = y$. We impose the side conditions that the metric g is complete (and homogeneous), that the subgroup of isometries fixing the point x is compact, and that $\mathfrak{I}(x)$ is maximal with respect to these properties. In principle, for each dimension m, we can describe the geometries arising. Note that the dimension of the isometry group $\mathfrak{I}(X) \leqslant \frac{m(m+1)}{2}$ and that $\mathfrak{I}(X)$ is compact if X is compact. In dimension three for example we consider Lie groups of dimension bounded by six, which at least provides a hint as to why our list above is exhaustive.

We have already mentioned the conjecture that 3-manifolds are built up from 'geometric pieces'. In higher dimensions this is known to be false, and can already be seen by looking at simply-connected examples. We conclude this subsection by introducing some of these using a different kind of construction.

In dimension five we need an invariant obtained from the second Stiefel-Whitney class, the class that measures the obstruction to the trivialisation of the tangent bundle restricted to the 2-skeleton or, equivalently, to the union of the 0-, 1- and 2-handles in a handle decomposition. (See section 7.7.) It may be regarded as a homomorphism $w : H_2(M, \mathbb{Z}) \longrightarrow \mathbb{Z}_2$ and we can write the abelian group $H_2(M, \mathbb{Z})$ as a direct sum of the maximal number of cyclic subgroups and then rechoose them so that w is non-zero on at most one summand. The order of this summand will be 2^i for $0 \leqslant i \leqslant \infty$. Then that $i \equiv i(M)$ is the invariant of M that we require and we have the following classification theorem.

Theorem 3.4.4. *Two simply connected 5-manifolds are diffeomorphic if and only if they have isomorphic second homology groups (with integer coefficients) and the same invariant $i(M)$ associated with the second Stiefel-Whitney class that we describe above.*

Any such manifold can be decomposed as a connected sum one manifold X_j, $-1 \leqslant j \leqslant \infty$, with possible extra summands from a second family of manifolds M_k, $1 < k \leqslant \infty$. The decomposition

$$X_j \# M_{k_1} \# M_{k_2} \# \cdots \# M_{k_s}$$

is unique provided that each k_i strictly divides k_{i+1} (or $k_{i+1} = \infty$). ■

To simplify the uniqueness statement we have taken $X_0 = \mathbf{S}^5$, however for discussion purposes it is convenient also to have $M_1 = \mathbf{S}^5$. Then, if we restrict attention to manifolds such that the structural group of the tangent bundle reduces to $U_2 \oplus 1$, that is, weakly almost complex manifolds, the only manifolds needed as summands are the M_k and X_∞.

The manifolds M_k, including $k = 1$, are such that $i(M_k) = 0$ and each bounds a manifold with trivialisable tangent bundle. The second homology group $H_2(M_k, \mathbb{Z})$ is isomorphic with $\mathbb{Z}/k \times \mathbb{Z}/k$, except for M_1 and $M_\infty = \mathbf{S}^2 \times \mathbf{S}^3$. The manifolds X_j similarly have $H_2(X_j, \mathbb{Z}) \cong \mathbb{Z}/2^j \times \mathbb{Z}/2^j$ except for X_0, X_{-1}, known as the Wu manifold, with $H_2(X_{-1}, \mathbb{Z}) \cong \mathbb{Z}/2$, and X_∞, which is diffeomorphic to a, actually the, non-trivial $\mathbf{S}^3$-bundle over $\mathbf{S}^2$.

All the summands are constructed by suitably gluing together two copies of boundary connected pairs, or singletons in the extreme cases, of disc bundles over $\mathbf{S}^2$, trivial bundles for the M_k and non-trivial ones for the X_j. See Barden

(1965) for full details. However the manifolds M_k admit an alternate beautiful description as 'Brieskorn varieties'.

Definition 3.4.5. Let $\underline{a} = (a_0, a_1, \cdots, a_n)$ be a sequence of $n+1$ natural numbers. The Brieskorn variety

$$V^{2n-1}(\underline{a}) = \{\underline{z} \in \mathbb{C}^{n+1} \big| z_0^{a_0} + \cdots + z_n^{a_n} = 0 \text{ and } z_0\overline{z}_0 + \cdots + z_n\overline{z}_n = 1\}$$

$V^{2n-1}(\underline{a})$ has the structure of $(2n-1)$-dimensional smooth manifold, which bounds a manifold with trivialisable tangent bundle. Its algebraic invariants, in particular its homology groups, can be calculated from the exponents a_i. See Hirzebruch and Mayer (1968) or Orlik and Wagreich (1975). With $n=3$ we have

$$M_k^5 = V^5(k,3,3,3)(3 \nmid k) \qquad \text{and} \qquad 3M_{3^t}^5 = V^5(3^t,4,4,4).$$

Hence, with some problems at the prime 3, weakly almost complex, simply-connected 5-manifolds are obtained using the connected sum construction from an easily described family of irreducibles.

We conclude our list of examples with one further class of Brieskorn varieties, illustrating the exotic differentiable structures which can arise on spheres.

Example 3.4.6. $V^7(3, 6k-1, 2, 2, 2)$ is homeomorphic to $\mathbf{S}^7$. For $k = 1, 2, \cdots, 28$ we obtain all possible smooth homotopy 7-spheres.

For a proof of this see Hirzebruch and Mayer (1968). The differential structures are distinguished by the algebraic invariants of the manifolds which they naturally bound in the Brieskorn construction. Note that each of them embeds in $\mathbf{S}^9$, that is in a sphere *two* dimensions greater than their own. The exoticity of their smooth structure requires such a jump in dimension.

3.5. Exercises for Chapter 3.

3.1. Show that the construction of the bundle E in Theorem 3.1.8 is unique up to the weak equivalence defined after the proof of that theorem.

3.2. Show that the regular octagon of Figure 3.1 exists in the disc model of the hyperbolic plane.

CHAPTER 4

DIFFERENTIAL FORMS AND INTEGRATION

Whereas assessing the differentiability of maps between manifolds was straightforward, granted our definition of the manifolds themselves, constructing a derivative for such maps took us a little more effort. Generalising the reverse operation of integration to manifolds will be even more demanding and we shall only achieve it in a limited context, equivalent to finding the definite integral of a real-valued function on a manifold, and even that will take us the whole of this chapter.

The first stage of the definition is to focus on what exactly it is that we are able to integrate. Classically we integrate over intervals of the real line or, more generally, domains in a Euclidean space and what we appear to be integrating are functions on those domains. However, even on the real line we commonly write the integrand as $f(x)\,\mathrm{d}x$ rather than just $f(x)$ and in an m-dimensional space we would write $f(x_1, \ldots, x_m)\,\mathrm{d}x_1 \ldots \mathrm{d}x_m$. In elementary presentations the $\mathrm{d}x$ is often treated as an optional extra. It is, however, a first clue to what we are in fact integrating: namely a '1-form' on $\mathbb{R}^1$ and an 'm-form' on $\mathbb{R}^m$. An important property of such forms that we shall discover, once we have defined them carefully, is that the rules for expressing them in different coordinate systems correspond precisely to the rules for changes of variables in the classical Riemann integrals. Further insight into the nature of our integrands can be gained from the definition of integrals over surfaces in $\mathbb{R}^3$. There, in addition to a function defined on the surface, we require a normal vector to that surface at each point. Typically this normal is produced as the cross product of two tangent vectors at a given point of the surface and then reduced to unit length by dividing by its norm. Thus, locally at least, the integrand – a two form – can be thought of as a function times the cross product of two vector fields on the surface.

Unlike the other basic concepts that we require to build up a theory of manifolds, the necessary linear, or more precisely multilinear, algebra that we now need is not often covered in undergraduate courses. This is the so-called *exterior algebra* of alternating multilinear functions on a vector space and we therefore give the definitions and summarise the main results that we need in the first section of the chapter. We are then, in section 2, able to define k-forms on an m-manifold using simultaneously the exterior algebra on all the tangent spaces to the manifold. From the results for the exterior algebra on a single vector space we deduce the various rules for the manipulation of forms that we shall require, and take for granted, in the remainder of the book.

However we are still not quite there. A second problem is hinted at in the definition of a surface integral: in which order do we take the cross-product of our tangent vectors? If we reverse the order we change the sign of the normal and hence also of the resulting integral. So we need to restrict our coordinate charts in some way that will tell us which order to take. That is achieved by an *orientation* of the manifold for which we give two equivalent definitions in section 3. Then, at last, in section 4 we are able to describe a well-defined definite integral of an m-form over an oriented m-manifold. At which point we close the chapter and pause for breath.

4.1. Forms on Vector Spaces.

We generalise 'linear forms', the elements of $V^* = \mathcal{L}(V, \mathbb{R})$, to define

$$\Lambda^k(V^*) = \{\text{alternating } k\text{-linear functions of } V\}$$

where

$$\mu : \underbrace{V \times \cdots \times V}_{k} \to \mathbb{R}$$

is *k-linear* if, when all but one argument is fixed, it is linear in the remaining one and it is *alternating* if it is zero whenever two arguments are equal. We shall see that

$$\dim(\Lambda^k(V^*)) = \begin{cases} \begin{pmatrix} n \\ k \end{pmatrix} & \text{if } 0 \leqslant k \leqslant n = \dim(V), \\ 0 & \text{otherwise.} \end{cases}$$

So, writing $\Lambda^*(V^*) = \bigoplus_{k=0}^{n} \Lambda^k(V^*)$, where $\Lambda^0(V^*) \cong \mathbb{R}$, we deduce that the dimension of $\Lambda^*(V^*)$ is 2^n. We call $\Lambda^*(V^*)$ a *graded vector space* with $\Lambda^k(V^*)$ in *degree k*. In fact it is a *graded algebra,* since we have the following product, denoted by '$\wedge$', and referred to as the *wedge product* or *exterior product*: for $\omega \in \Lambda^k(V^*)$ and $\theta \in \Lambda^l(V^*)$ we define $\omega \wedge \theta \in \Lambda^{k+l}(V^*)$ by

$$\begin{aligned}(\omega \wedge \theta)(v_1, \cdots, v_{k+l}) &= \frac{1}{k!l!} \sum_{\sigma \in \mathcal{S}_{k+l}} \epsilon_\sigma \omega(v_{\sigma_1}, \cdots, v_{\sigma_k})\, \theta(v_{\sigma_{k+1}}, \cdots, v_{\sigma_{k+l}}) \\ &= \sum_{(k,l)-\text{shuffles}} \epsilon_\sigma \omega(v_{\sigma_1}, \cdots, v_{\sigma_k})\, \theta(v_{\sigma_{k+1}}, \cdots, v_{\sigma_{k+l}})\end{aligned}$$

where $\mathcal{S}_{k+l}$ is the symmetric group of all permutations of the $k+l$ suffices and ϵ_σ is the sign, ± 1, of the permutations σ. This summation is necessary to ensure that the resulting function is still alternating: when the two equal arguments appear in the same factor that is already guaranteed, but when they are in different factors, one an argument for ω and one for θ, we need the similar term in which they, and only they, are exchanged and which therefore appears in the sum with the opposite sign. However the full summation has considerable redundancy: for any given permutation in the sum, any further permutation just of the arguments of ω and of the arguments of θ that does not mix the two sets of arguments is un-necessary. This is taken care of by the external factor $\frac{1}{k!l!}$ in the first expression and by the restriction of the permutations in the second. The permutations occurring in the second expression are referred to as (k, l)-*shuffles* since such a permutation is what results when a pack of $k + l$ cards is 'cut' into two subpacks of k and l cards which are then spliced together, preserving the order of the cards in each sub-pack.

Note that $1 \in \mathbb{R} = \Lambda^0(V^*)$ is a multiplicative identity for this algebra. We also note the following elementary properties: the algebra is associative and graded commutative, where the latter means $\theta \wedge \omega = (-1)^{\deg(\omega)\deg(\theta)} \omega \wedge \theta$; the wedge product is bilinear so, for example,

$$\theta \wedge (\lambda_1 \omega_1 + \lambda_2 \omega_2) = \lambda_1 \theta \wedge \omega_1 + \lambda_2 \theta \wedge \omega_2.$$

Proposition 4.1.1. *If* $\phi_1, \cdots, \phi_k \in V^*$ *and* $v_1, \cdots, v_k \in V$, *then*

$$(\phi_1 \wedge \cdots \wedge \phi_k)(v_1, \cdots, v_k) = \det(\phi_i(v_j)).$$

Proof. The result is true, trivially, for $k = 1$. So, for a proof by induction, we may assume it for $k-1$. Then

$$\begin{aligned}
&(\phi_1 \wedge (\phi_2 \wedge \cdots \wedge \phi_k))(v_1, \cdots, v_k) \\
&= \sum_{(1,k-1)\text{-shuffles } \sigma} \epsilon_\sigma \phi_1(v_{\sigma_1})(\phi_2 \wedge \cdots \wedge \phi_k)(v_{\sigma_2}, \cdots, v_{\sigma_k}) \\
&= \sum_{j=1}^{k} (-1)^{j-1} \phi_1(v_j) \det \big(\phi_l(v_m)\big)_{\hat{1}\hat{j}} \\
&= \det(\phi_i(v_j)),
\end{aligned}$$

where $\big(\phi_l(v_m)\big)_{\hat{1}\hat{j}}$ denotes the matrix $\big(\phi_l(v_m)\big)$ with the 1st row and jth column deleted, and the result follows since the penultimate line is just the formula for expansion of the determinant of the full matrix by its first row. ∎

Proposition 4.1.2. *If (e_i) is a basis of V and (ϵ^j) the dual basis of V^*, so that $\epsilon^j(e_i) = \delta_i^j$, then*

$$\{\epsilon^{i_1} \wedge \cdots \wedge \epsilon^{i_k} \mid 1 \leqslant i_1 < i_2 < \cdots < i_k \leqslant n\}$$

is a basis of $\Lambda^k(V^*)$.

Proof. For linear independence suppose

$$\sum_{i_1 < \cdots < i_k} a_{i_1 \cdots i_k} \epsilon^{i_1} \wedge \cdots \wedge \epsilon^{i_k} = 0.$$

Then evaluating on the k-tuple $(e_{j_1}, \cdots, e_{j_k})$ where $j_1 < \cdots < j_k$, we get $a_{j_1 \cdots j_k} = 0$. Hence all coefficients vanish, and the stated forms are linearly independent. To see that they span $\Lambda^k(V^*)$, let ω be an arbitrary element of $\Lambda^k(V^*)$ and define

$$\theta = \sum_{i_1 < \cdots < i_k} \omega(e_{i_1}, \cdots, e_{i_k})\, \epsilon^{i_1} \wedge \cdots \wedge \epsilon^{i_k}.$$

Then θ is in $\Lambda^k(V^*)$ and $\theta(e_{i_1}, \cdots, e_{i_k}) = \omega(e_{i_1}, \cdots, e_{i_k})$ for all $i_1, \cdots, i_k$. Hence, by the multilinearity of both θ and ω, θ and ω take the same values on all k-tuples of vectors $(v_1, \cdots, v_k)$. So ω is the stated linear combination θ of the given elements of $\Lambda^k(V^*)$. ∎

Notation. It is convenient to write I for the 'multi-index' $(i_1, \cdots, i_k)$, where $1 \leqslant i_1 < \cdots < i_k \leqslant n$; ϵ^I for the repeated wedge product $\epsilon^{i_1} \wedge \cdots \wedge \epsilon^{i_k}$;

and e_I, for example, for the k-tuple of vectors $(e_{i_1}, \cdots, e_{i_k})$ and v_I for a k-tuple of general vectors $(v_{i_1}, \cdots, v_{i_k})$. Stretching the notation slightly further, we shall also write $\alpha(v_I)$ for the k-tuple $(\alpha(v_{i_1}), \cdots, \alpha(v_{i_k}))$, where α is an endomorphism of V.

Then we may write $\omega = \sum_I a_I \, \epsilon^I$, $\theta = \sum_J b_J \, \epsilon^J$ so that, by the bilinearity of $\wedge$,

$$\omega \wedge \theta = \sum_{I,J} a_I \, b_J \, \epsilon^I \wedge \epsilon^J .$$

We also recall that a linear map $\alpha : V \to W$ determines a dual linear map $\alpha^* : V^* \leftarrow W^*$ by $\alpha^*(\omega)(v) = \omega(\alpha(v))$. We similarly define the linear map

$$\alpha^* : \Lambda^k(V^*) \longleftarrow \Lambda^k(W^*)$$

by $\alpha^*(\omega)(v_1, \cdots, v_k) = \omega(\alpha(v_1), \cdots, \alpha(v_k))$.

4.2. Forms on Manifolds.

From the tangent space $\tau_p(M)$ at p we may derive, as well as the cotangent space $\tau_p^*(M)$, the exterior powers $\Lambda^k(\tau_p^*(M))$ or 'space of k-forms at p', which we shall denote more concisely by $\lambda_p^k(M)$. Once again a coordinate chart (ϕ, U) induces isomorphisms $\phi_p^* : \lambda_p^k(M) \stackrel{\cong}{\leftarrow} \lambda_{\phi(p)}^k(\mathbb{R}^m)$ and thus bijections linear on each fibre

$$\Phi^* : \lambda^k(U) = \bigcup_{p \in U} \lambda_p^k(M) \stackrel{\cong}{\longleftarrow} \lambda^k(\phi(U)) = \bigcup_{q \in \phi(U)} \lambda_q^k(\mathbb{R}^m).$$

Then the kth exterior powers of the parallel translation isomorphisms produce further bijections from $\lambda^k(\phi(U))$ to $\phi(U) \times \Lambda^k(\mathbb{R}^m) \cong \phi(U) \times \mathbb{R}^{\binom{m}{k}}$. As before these induce topological and differential structures on $\lambda^k(M)$ since the structures from different charts differ by the coordinate translation $\theta_{\alpha\beta} \times (D\theta_{\alpha\beta}^{-1})^*$.

Then $\lambda^k(M)$ is called the *bundle of k-forms on M*, though strictly a *k-form on $U \subseteq M$* is a cross-section ω of $\lambda^k(M)$ defined on U, that is, a smooth map $\omega : U \to \lambda^k(M)$ such that $\pi \circ \omega = \mathrm{id}_U$, where π is the usual projection mapping $\lambda_p^k(M)$ to $p \in M$.

We write $\Omega^k(U)$ for the vector space, in fact it is a $\mathcal{C}^\infty(U)$-module, of all k-forms on U.

We usually refer to the 1-form $\phi^*(d\pi_i)$ as dx^i on the domain U of ϕ. Then the $dx^i(p)$ form a basis of $\lambda_p^1(U)$ and $dx^{i_1}(p) \wedge \cdots \wedge dx^{i_k}(p)$, where $1 \leqslant i_1 < \cdots < i_k \leqslant m$, form a basis of $\lambda_p^k(M)$.

So locally any form on U is $\sum_I a_I \, dx^I$, in the notation introduced in Section 4.1. In fact this looks very similar to the notation for a form on a vector space, but now the a_I are functions on U. The form is smooth if and only if the a_I are smooth functions. There is an alternative characterisation of the smoothness of forms, which is often convenient: for k vector fields $X_1, \cdots, X_k$ and a k-form ω we may define the evaluation function $\omega(X_1, \cdots, X_k)$. Then ω is smooth if and only if the function $\omega(X_1, \cdots, X_k)$ is smooth whenever all the vector fields $X_1, \cdots, X_k$ are smooth.

Definition 4.2.1. If $f : M \to N$ is a smooth map, the homomorphism from $\lambda_{f(p)}^k(N)$ to $\lambda_p^k(M)$ induced by the derivative $f_*(p)$ transforms a k-form $\omega : N \to \lambda^k(N)$ on N into a k-form $f^*(\omega)$ on M called the *pull-back of* ω *by* f. In detail $f^*(\omega)$ is given by the formula:

$$f^*(\omega)(p)(v_1, \cdots, v_k) = \omega(f(p))(f_*(v_1), \cdots, f_*(v_k)).$$

For '0-forms', that is smooth functions $g \in \mathcal{C}^\infty(N)$, we define $f^*(g)$ to be $g \circ f \in \mathcal{C}^\infty(M)$.

These pull-backs will play an important rôle in our account of integration on manifolds, and we shall need the following facts in our calculations.

Proposition 4.2.2. 1*a*) $f^*(\lambda_1\omega_1 + \lambda_2\omega_2) = \lambda_1 f^*(\omega_1) + \lambda_2 f^*(\omega_2)$, *where the* ω_i *are* k*-forms and the* λ_i *real.*

1*b*) $f^*(\omega_1 \wedge \omega_2) = f^*(\omega_1) \wedge f^*(\omega_2)$. *In particular, when* g *is a smooth function on* N, $f^*(g\omega) = f^*(g)f^*(\omega) = (g \circ f)f^*(\omega)$.

2) $(g \circ f)^*(\omega) = f^*(g^*(\omega))$ *where* g *is a smooth map from* N *to a manifold* P *and* ω *is a* k*-form on* P. *The pull-back of the identity is the identity.*

Remark 4.2.3. The results (1) say that f^* is an algebra homomorphism. The statements (2) say that pull-back is a contravariant functor.

Proof. (1*a*) This follows from the pointwise definition of the linear combination of forms. The scalar multiplication by λ is covered by case (1*b*) with ω_1 the constant function $g(x) = \lambda$. Then, writing v_I for a k-tuple $(v_1, \cdots, v_k)$ of vectors we have:

$$\begin{aligned} f^*(\omega_1+\omega_2)(p)(v_I) &= (\omega_1+\omega_2)(f(p))(f_*(v_I)) \\ &= \omega_1(f(p))(f_*(v_I)) + \omega_2(f(p))(f_*(v_I)) \\ &= f^*(\omega_1)(p)(v_I) + f^*(\omega_2)(p)(v_I) \\ &= (f^*(\omega_1)(p) + f^*(\omega_2)(p))(v_I) \\ &= (f^*(\omega_1) + f^*(\omega_2))(p)(v_I), \end{aligned}$$

where $f_*(v_I)$ denotes the k-tuple $(f_*((v_1), \cdots, f_*(v_k))$. Since this is true at all points p and for all k-tuples of vectors v_I, we have $f^*(\omega_1+\omega_2) = f^*(\omega_1) + f^*(\omega_2)$ as required.

(1b) Since $(\omega_1 \wedge \omega_2)(p) = \omega_1(p) \wedge \omega_2(p)$ it suffices to work at p, explicit mention of which we shall suppress. Then

$$\begin{aligned} f^*(\omega_1\wedge\omega_2)(v_1\cdots,v_{k+l}) &= (\omega_1\wedge\omega_2)(f_*v_1\cdots,f_*v_{k+l}) \\ = \sum_{\text{shuffles } \sigma} & \epsilon_\sigma \omega_1(f_*(v_{\sigma(1)})\cdots,f_*(v_{\sigma(k)}))\,\omega_2(f_*(v_{\sigma(k+1)})\cdots,f_*(v_{\sigma(k+l)})) \\ = \sum_{\text{shuffles } \sigma} & \epsilon_\sigma(f^*(\omega_1)(v_{\sigma(1)}\cdots,v_{\sigma(k)})f^*(\omega_2)(v_{\sigma(k+1)}\cdots,v_{\sigma(k+l)}) \\ =&(f^*\omega_1\wedge f^*\omega_2)(v_1\cdots,v_{k+l}) \end{aligned}$$

for all $(v_1\cdots,v_{k+l})$. Hence $f^*(\omega_1\wedge\omega_2) = f^*(\omega_1)\wedge f^*(\omega_2)$.

(2) This follows from the chain rule, $(g\circ f)_* = g_*\circ f_*$. Using the notation v_I and $f_*(v_I)$ again as above, and also suppressing the the reference to the points p, $f(p)$ and $g(f(p))$ at which the various vectors act we see:

$$\begin{aligned} (g\circ f)^*(\omega)(v_I) &= \omega((g\circ f)_*(v_I)) = \omega(g_*(f_*(v_I))) \\ &= (g^*\omega)(f_*(v_I)) = (f^*(g^*\omega))(v_I). \end{aligned}$$

Again this is valid for all v_I, so we deduce $(g\circ f)^*(\omega) = f^*(g^*(\omega))$. ■

We shall also need the following explicit formula for the pull-backs of forms on subsets of Euclidean spaces, in particular for the pull-back by a diffeomorphism of an m-form on $\mathbb{R}^m$.

Proposition 4.2.4. 1) *If V and W are open in $\mathbb{R}^m$ and $\mathbb{R}^n$ with coordinates (x_i) and (y_j) respectively and $\theta : W \to V$ is a smooth map, then $\theta^*(dx^i) = \frac{\partial\theta_i}{\partial y_j}dy^j$.*

2) *If θ is a diffeomorphism (so that $m = n$) and $\omega = f\, dx^1 \wedge \cdots \wedge dx^m$ is a (general) m-form on V, then*

$$\theta^*(\omega) = (f \circ \theta)(\det(J(\theta)))(y)\, dy^1 \wedge \cdots \wedge dy^m.$$

Proof. If $(\theta^*(dx^i)) = f_j\, dy^j$, then $f_j = (\theta^*(dx^i))\left(\frac{\partial}{\partial y_j}\right)$. However,

$$(\theta^*(dx^i))\left(\frac{\partial}{\partial y_j}\right) = dx^i\left(\theta_*\frac{\partial}{\partial y_j}\right) = dx^i\left(\frac{\partial \theta_k}{\partial y_j}\frac{\partial}{\partial x_k}\right) = \frac{\partial \theta_i}{\partial y_j},$$

which establishes (1).

Then, by part (1b) of Proposition 4.2.2,

$$\theta^*(\omega) = \theta^*(f)\,\theta^*(dx^1) \wedge \theta^*(dx^2) \wedge \cdots \wedge \theta^*(dx^m).$$

So, by Proposition 4.1.1,

$$\theta^*(\omega)(y)\left(\frac{\partial}{\partial y_1}, \cdots, \frac{\partial}{\partial y_m}\right) = (f \circ \theta(y)) \cdot \det\left((\theta^*(dx^i))\left(\frac{\partial}{\partial y_j}\right)\right).$$

But then, using part (1),

$$\theta^*(\omega)(y)\left(\frac{\partial}{\partial y_1}, \cdots, \frac{\partial}{\partial y_m}\right) = (f \circ \theta(y)) \cdot \det(J(\theta)(y))$$

and $\theta^*(\omega) = (f \circ \theta)(\det(J(\theta))\, dy^1 \wedge \cdots \wedge dy^m$, since $dy^1 \wedge \cdots \wedge dy^m$ is the only basis vector in $\lambda^m(\mathbb{R}^m)$ and its coefficient is determined by its value on $\left(\frac{\partial}{\partial y_1}, \cdots, \frac{\partial}{\partial y_m}\right)$. ■

4.3. The Orientation of Manifolds.

The m-forms we have just constructed are what we shall be integrating over manifolds. However, before we can do so, we need a refinement of the differential structure. We need it to be oriented, which means that the derivatives of all its coordinate transformations have Jacobian matrices with positive determinant. We have already given the following definition in Chapter 3 (Definition 3.2.6). We repeat it here for those readers who may have omitted that chapter.

Definition 4.3.1. An atlas $\mathcal{A}$ on M is *oriented* and is called *an orientation of* M if, for all charts (ϕ_α, U_α) and (ϕ_β, U_β) in $\mathcal{A}$ with coordinate transformation $\theta_{\alpha\beta}$, $\det(J(\theta_{\alpha\beta}))$ is positive on its domain $\phi_\alpha(U_\alpha \cap U_\beta)$. An *oriented* (smooth) *manifold* is $(M, \mathcal{A})$, a topological manifold M together with a (smooth) oriented atlas $\mathcal{A}$ on M. We say that a (smooth) manifold M is *orientable* if it possesses an oriented atlas.

Remarks 4.3.2. 1) As before, more precisely, an *orientation* of a topological manifold is a *maximal* oriented atlas, and an oriented manifold is a topological manifold together with an orientation.

2) If $\rho : \mathbb{R}^m \to \mathbb{R}^m$ is a linear isomorphism with $\det(\rho) = -1$ then, for each orientation $\mathcal{A}$, there is the *opposite orientation* $-\mathcal{A} = \{(\rho \circ \phi, U) \mid (\phi, U) \in \mathcal{A}\}$.

Definition 4.3.3. A never-zero m-form on M^m is called a *volume form*.

Remark 4.3.4. If volume forms exist they are not unique, for if ω is a volume form and f a never-zero function on M then $f\omega$ is also a volume form. Never-zero functions of course always exist, for example the non-zero constant functions.

Proposition 4.3.5. *A compact manifold M is orientable if and only if there exists a volume form on M.*

Proof. Suppose that M has a volume form ω. A chart (ϕ, U) will transfer this to a form, $(\phi^{-1})^*(\omega|_U)$, on the coordinate space $V \subseteq \mathbb{R}^m$ where, being an m-form, it will be given by $a(x)dx^1 \wedge \cdots \wedge dx^m$ for some never-zero function a on V. By reversing the x^1 axis if necessary, we may assume that $a > 0$. Then, in the coordinate space of another chart, ω is represented by $b(y)\, dy^1 \wedge \cdots \wedge dy^m$ again with $b(y) > 0$. But, if θ is the coordinate transformation between the two, then $b(y) = a(\theta(y)) \det(J(\theta))$ by Proposition 4.2.4. So, with these choices of local coordinates, $\det(J(\theta)) = b(y)/a(\theta(y)) > 0$.

Conversely let (f_i) be a partition of unity subordinate to a cover of M by neighbourhoods (U_i) of charts (ϕ_i, U_i) belonging to an oriented atlas. For coordinates $(x_1, \cdots, x_m)$ on the coordinate space $\phi_i(U_i)$ define an m-form on M by

$$\omega_i(p) = \begin{cases} f_i(p)\phi_i^*(dx^1 \wedge \cdots \wedge dx^m)(p) & \text{if } p \in U_i, \\ 0 & \text{if } p \notin U_i. \end{cases}$$

As usual this is well-defined and smooth on M. Then $\omega = \sum \omega_i$ is well-defined and given on U_k by

$$\begin{aligned}
&\sum_i f_i(p)\phi_k^*((\theta_{ik})^*(dx^1 \wedge \cdots \wedge dx^m)) \qquad \text{since } {\phi_i}^* = (\theta_{ik} \circ \phi_k)^*, \\
&= \sum_i f_i(p)\phi_k^*\big(\det(J(\theta_{ik}))dy^1 \wedge \cdots \wedge dy^m\big) \\
&= \left(\sum_i f_i(p)\Big(\det\big(J(\theta_{ik})\big)(\phi_k(p))\Big)\right)\phi_k^*(dy^1 \wedge \cdots \wedge dy^m).
\end{aligned}$$

This is never-zero since all coefficients are non-negative and at each point at least one is strictly positive. ∎

Note that the hypothesis of compactness may be removed since it was only required in the above proof to ensure the existence of partitions of unity.

4.4. Integration of m-forms on Oriented m-manifolds.

We are now ready for the main concept of this chapter, that of the integral of an m-form over an oriented m-manifold. As usual we work first on a coordinate neighbourhood, check what happens under changes of coordinates and then integrate over the entire manifold. We restrict attention to compact manifolds since we shall use partitions of unity which we have only obtained for such manifolds. However since partitions of unity exist for non-compact, but still second countable or paracompact, manifolds so too do integrals of top-dimensional forms.

So we consider an m-form ω on a compact oriented m-dimensional manifold M with the support of ω, the closure of the set of points at which it is non-zero, contained in the coordinate neighbourhood U_α of a chart ϕ_α in the oriented atlas on M. Let $({\phi_\alpha}^{-1})^*(\omega) = a_\alpha(x)\, dx^1 \wedge \cdots \wedge dx^m$. Then we define $\int_M \omega \equiv \int_{U_\alpha} \omega$ to be the multiple Riemann integral over $\phi_\alpha(U_\alpha) \subseteq \mathbb{R}^m$ of the function a_α:

$$\int_M \omega = \int_{\phi_\alpha(U_\alpha)} a_\alpha(x)\, dx^1 \cdots dx^m.$$

Then, if the support of ω is also contained in the coordinate neighbourhood U_β of another chart ϕ_β in the same orientation of M as ϕ_α, we have $\phi_\alpha = \theta_{\alpha\beta} \circ \phi_\beta$

where, as usual, we restict the domains as necessary for this to make sense. So

$$\begin{aligned}(\phi_\alpha^{-1})^*(\omega) &= (\phi_\beta^{-1}\circ\theta_{\alpha\beta}^{-1})^*(\omega)\\ &= (\theta_{\alpha\beta}^{-1})^*\circ(\phi_\beta^{-1})^*(\omega) = (\theta_{\beta\alpha})^*\circ(\phi_\beta^{-1})^*(\omega)\\ &= (\theta_{\beta\alpha})^*(a_\beta\,dy^1\wedge\cdots\wedge dy^m)\\ &= (a_\beta\circ\theta_{\beta\alpha})\det(J(\theta_{\beta\alpha}))\,dx^1\wedge\cdots\wedge dx^m.\end{aligned}$$

So $a_\alpha(x) = (a_\beta\circ\theta_{\beta\alpha})(x)\det(J(\theta_{\beta\alpha}))(x)$ and thus, writing $U_{\alpha\beta}$ for $U_\alpha\cap U_\beta$,

$$\begin{aligned}\int_{U_\alpha}\omega = \int_{U_{\alpha\beta}}\omega &= \int_{\phi_\alpha(U_{\alpha\beta})} a_\alpha(x)\,dx^1\cdots dx^m\\ &= \int_{\theta_{\alpha\beta}(\phi_\beta(U_{\alpha\beta}))} a_\alpha(x)\,dx^1\cdots dx^m\\ &= \int_{\phi_\beta(U_{\alpha\beta})} a_\alpha\circ\theta_{\alpha\beta}(y)\det(J(\theta_{\alpha\beta}))\,dy^1\cdots dy^m\end{aligned}$$

by the change of variable formula for multiple integrals in $\mathbb{R}^m$. Usually this would require $|\det(J(\theta_{\alpha\beta}))|$ but here, since both charts belong to the same orientation, we already know that $\det(J(\theta_{\alpha\beta}))>0$. This is precisely why we need our manifolds to be oriented.

Thus

$$\int_{U_\alpha}\omega = \int_{\phi_\beta(U_{\alpha\beta})} a_\beta(y)\,dy^1\wedge\cdots\wedge dy^m = \int_M\omega,$$

where the latter is defined using the chart (ϕ_β, U_β).

Thus this definition is independent of the choice of coordinate neighbourhood containing the support of ω and we may now generalise.

Definition 4.4.1. Let ω be an m-form on an oriented manifold M^m and $\{g_i \mid i = 1,\cdots,k\}$ a partition of unity subordinate to a covering by coordinate neighbourhoods $\{U_i\}$ from the orientation of M. Then we define $\int_M\omega = \sum_{i=1}^{k}\int_M g_i\,\omega$, where $\int_M g_i\,\omega$ is defined as above, using the fact that $g_i\,\omega$ is supported on U_i.

Lemma 4.4.2. *This is well-defined.*

Proof. Let $\{h_j \mid j = 1,\cdots,l\}$ be another partition of unity subordinate to $\{V_j\}$ on M. Then $\{g_ih_j\}$ is a partition of unity subordinate to $\{U_i\cap V_j\}$ on

M and

$$\sum_{i=1}^{k}\int_M g_i\,\omega = \sum_{i=1}^{k}\int_M \sum_{j=1}^{l} h_j\,g_i\,\omega = \sum_{i=1}^{k}\sum_{j=1}^{l}\int_M h_j\,g_i\,\omega$$
$$= \sum_{j=1}^{l}\int_M \sum_{i=1}^{k} h_j\,g_i\,\omega = \sum_{j=1}^{l}\int_M h_j\left(\sum_{i=1}^{k} g_i\,\omega\right)$$
$$= \sum_{j=1}^{l}\int_M h_j\,\omega.$$

■

4.5. Exercises for Chapter 4.

4.1. Prove that for a real valued multilinear form μ on a real vector space V the following are equivalent.
   (*i*) $\mu = 0$ whenever two arguments agree.
   (*ii*) μ changes sign if two arguments are interchanged leaving all the other arguments fixed.
   (*iii*) For any permutation σ of the k arguments

$$\mu(v_1,\cdots,v_k) = \epsilon_\sigma \mu(v_{\sigma(1)},\cdots,v_{\sigma(k)})$$

   where ϵ_σ is the sign of σ.

4.2. Let α be a linear map of the m-dimensional vector space V into itself, and let $\omega \in \wedge^m(V^*)$. Calculate $\alpha^*(\omega)$.

4.3. Show that, for 1-forms φ_i, $\varphi_1\wedge\cdots\wedge\varphi_k = 0$ if and only if the φ_i are linearly dependent. If they are linearly independent, prove that $\varphi_1\wedge\cdots\wedge\varphi_k = \psi_1\wedge\cdots\wedge\psi_k$ for 1-forms ψ_j if and only if $\varphi_i = \sum_j a_{ij}\psi_j$ with $\det(a_{ij}) = 1$.

4.4. A non-zero k-form φ is called *decomposable* if $\varphi = \varphi_1\wedge\cdots\wedge\varphi_k$ where the φ_i are all 1-forms. Show that, if $\dim(V) < 4$, then every non-zero element of $\wedge^2(V^*)$ is decomposable. Give a counter-example for dimension four.

4.5. The Hodge star isomorphism $*$ from $\Omega^k(\mathbb{R}^m)$ to $\Omega^{m-k}(\mathbb{R}^m)$ is defined by mapping the basic k-form $dx^{i_1}\wedge\cdots\wedge dx^{i_k}$ to $\epsilon_\sigma dx^{j_1}\wedge\cdots\wedge dx^{j_{m-k}}$ where $i_1<\cdots<i_k$, $j_1<\cdots<j_{m-k}$ and $(i_1,i_2,\cdots,i_k,j_1,\cdots,j_{m-k})$ is the permutation σ of $(1,2,\cdots,m)$. Let

$$\omega = a_{12}dx^1\wedge dx^2 + a_{13}dx^1\wedge dx^3 + a_{23}dx^2\wedge dx^3 \in \Omega^2(\mathbb{R}^3).$$

   Calculate $*\omega$. What is $*\omega$ if $\omega\in\Omega^2(\mathbb{R}^4)$?

CHAPTER 5

THE EXTERIOR DERIVATIVE

In introducing the previous chapter we implicitly claimed that our integral of m-forms over oriented m-manifolds was, in some sense, the reverse operation to differentiation. Classically this duality is expressed by the fundamental theorem of the integral calculus:

$$\int_a^b f'(x)\,dx = f(b) - f(a).$$

The prime aim of this chapter is to prove Stokes' Theorem which is, in fact, the generalisation to our present context of that theorem.

The integrand $f'(x)\,dx$ may be re-written simply as df and the first task in this chapter is to generalise the map $f \mapsto df$ from functions to 1-forms, that is, to define the exterior derivative d which maps k-forms on a manifold to $(k+1)$-forms. As usual we do this first locally, for k-forms on a Euclidean space, in section 1 before giving two equivalent global definitions in section 2.

Looking again at the fundamental theorem above, the clue to the generalisation that we seek is to think of the expression $f(b) - f(a)$ as the evaluation of the function f on the oriented boundary of the interval $[a, b]$. Thus, in section 3, we shall extend our definition of manifolds to allow them to have boundaries, with an orientation of the manifold inducing a specific orientation of its boundary. That done, in the next section we state and prove Stokes' Theorem and give, by way of illustration, a few of its many applications.

In the final section we prove a technical result which, for want of better inspiration, we refer to as the 'bubbling' of top dimensional forms on a manifold. The application that we give of it here will have an important re-interpretation in the next chapter.

5.1. The Exterior Derivative on $\mathbb{R}^m$.

We recall that a general k-form ω on an open subset U of $\mathbb{R}^m$ may be written as $\sum_I a_I \, dx^I$ where I ranges over all strictly increasing multi-indices of length k. Since the a_I are functions on U, we have already defined da_I and shown it to be equal to $\frac{\partial a_I}{\partial x_i} dx^i$.

Definition 5.1.1. For a k-form $\omega = \sum_I a_I \, dx^I$ on an open subset U of $\mathbb{R}^m$ we define $d\omega$ to be $\sum_I da_I \wedge dx^I$.

In particular this definition implies that $d(dx^I) = d1 \wedge dx^I = 0$. For any general explicit expression of ω the corresponding expression for $d\omega$ requires simplification, of course. For example, if $m = 3$ and $\omega = a(x)\, dx^2 \wedge dx^3$, then $d\omega = \frac{\partial a}{\partial x_1} dx^1 \wedge dx^2 \wedge dx^3$ since the other two terms are multiples of the forms $dx^2 \wedge dx^2 \wedge dx^3$ and $dx^3 \wedge dx^2 \wedge dx^3$ which are both zero.

Proposition 5.1.2. 1) *d is $\mathbb{R}$-linear: $d(\lambda_1\omega_1 + \lambda_2\omega_2) = \lambda_1 d\omega_1 + \lambda_2 d\omega_2$ for real constants λ_i and k-forms ω_i.*

2) *$d(\omega_1 \wedge \omega_2) = d\omega_1 \wedge \omega_2 + (-1)^k \omega_1 \wedge d\omega_2$ for a k-form ω_1 and any form ω_2.*

3) *$d^2 = 0$.*

4) *$d(f^*\omega) = f^*(d\omega)$ where ω is a k-form on $U \subseteq \mathbb{R}^m$, $V \subseteq \mathbb{R}^n$ and $f : V \to U$ is differentiable.*

Remark 5.1.3. Note in particular when ω_1 is a 0-form, that is a differentiable function $f : \mathbb{R}^m \to \mathbb{R}^n$, property (2) says that $d(f\omega) = df \wedge \omega + f(d\omega)$. This should be compared and contrasted with the definition which relies on ω being expressed as a sum of the basic forms dx^I. Then the term $f_I\,(d(dx^I))$, which the general result would require, is zero since as already noted $d(dx^I) = 0$. Property (2) is in fact saying that d is a derivation of the graded algebra of forms on U, the grading accounting for the sign $(-1)^k$.

Proof. (1) Additivity is a straightforward check. That $d(\lambda\omega) = \lambda d(\omega)$ is the special case of property (2) where ω_1 is the constant function with value $\lambda \in \mathbb{R}$.

(2) When $\deg(\omega_1) = \deg(\omega_2) = 0$ this is just the product rule for differentials: $d(fg) = f(dg) + g(df)$, except that now we know what we are talking

about. In general, let $\omega_1 = \sum_I a_I\, dx^I$ and $\omega_2 = \sum_J b_J\, dx^J$. Then

$$\begin{aligned}
d(\omega_1 \wedge \omega_2) &= \sum_{I,J} d(a_I\, b_J) \wedge dx^I \wedge dx^J \\
&= \sum_{I,J} b_J\, da_I \wedge dx^I \wedge dx^J + \sum_{I,J} a_I\, db_J \wedge dx^I \wedge dx^J \\
&= \sum_J b_J\, d\omega_1 \wedge dx^J + (-1)^k \sum_{I,J} a_I\, dx^I \wedge db_J \wedge dx^J
\end{aligned}$$

which is $d\omega_1 \wedge \omega_2 + (-1)^k \omega_1 \wedge d\omega_2$, as required, the $(-1)^k$ factor coming from the fact that the 1-form db_J has been interchanged with the k-form dx^I.

(3) For a 0-form, or function, a we have

$$d(da) = d\left(\frac{\partial a}{\partial x_j} dx^j\right) = d\left(\frac{\partial a}{\partial x_j}\right) \wedge dx^j = \frac{\partial^2 a}{\partial x_i \partial x_j} dx^i \wedge dx^j = 0$$

since $\frac{\partial^2 a}{\partial x_i \partial x_j} = \frac{\partial^2 a}{\partial x_j \partial x_i}$ and $dx^i \wedge dx^j = -dx^j \wedge dx^i$. Then for a general k-form $\omega = \sum_I a_I\, dx^I$ we have $d\omega = \sum_I da_I \wedge dx^I$ so that, by property (2),

$$d^2\omega = \sum_I \left\{ d(da_I) \wedge dx^I - da_I \wedge d(dx^I) \right\}$$

which is zero since both the $d(da_I)$ and the $d(dx^I)$ are zero.

(4) Let $\omega = g$ be a 0-form on $\mathbb{R}^m$. Then

$$\begin{aligned}
f^*(dg) &= f^*\left(\frac{\partial g}{\partial x_i} dx^i\right) = \left(\frac{\partial g}{\partial x_i} \circ f\right) f^*(dx^i) = \left(\frac{\partial g}{\partial x_i} \circ f\right) \frac{\partial f_i}{\partial y_j}\, dy^j \\
&= \frac{\partial (g \circ f)}{\partial y_j}\, dy^j = d(g \circ f) = d(f^*(g)).
\end{aligned}$$

Then for the basic 1-form dx^i we have $f^*(dx^i) = \frac{\partial f_i}{\partial y_j} dy^j$ and so

$$d(f^*(dx^i)) = d\left(\frac{\partial f_i}{\partial y_j}\right) \wedge dy^j = \frac{\partial^2 f_i}{\partial y_k \partial y_j} dy^k \wedge dy^j = 0 = f^*(d(dx^i)),$$

by property (3). For a basic k-form dx^I, $f^*(d(dx^I)) = 0$ since $d(dx^I) = 0$. On the other hand, since, by Proposition 4.2.2, the pull-back f^* is an algebra homomorphism, we get

$$d(f^* dx^I) = d\left((f^* dx^{i_1}) \wedge (f^* dx^{i_2}) \wedge \cdots \wedge (f^* dx^{i_k})\right),$$

which is zero since, by property (2), it is a sum of terms each having a factor $d(f^*dx^i) = 0$. So for a general k-form $\omega = \sum_I a_I\, dx^I$ we have

$$d(f^*(\omega)) = d\sum_I f^*(a_I)f^*(dx^I) = \sum_I d(f^*(a_I)) \wedge f^*(dx^I)$$

by property (2), again using the fact that the pull-back f^* is an algebra homomorphism. So, by the result for 0-forms,

$$d(f^*(\omega)) = \sum_I f^*(da_I) \wedge f^*(dx^I) = f^* \sum_I da_I \wedge dx^I = f^*(d\omega). \qquad \blacksquare$$

5.2. The Exterior Derivative on Manifolds.

Having defined, and established the properties of, the exterior derivative of forms on $\mathbb{R}^n$, we now generalise it to manifolds. Recall that, given a chart (ϕ_α, U_α) on a manifold M, a differentiable k-form ω is represented locally on $V_\alpha = \phi_\alpha(U_\alpha)$ by $\omega_\alpha = (\phi_\alpha^{-1})^*(\omega) \in \Omega^k(\phi_\alpha(U_\alpha))$. Then the restriction of ω to U_α is $\phi_\alpha^*(\omega_\alpha)$, since ${\phi_\alpha}^{-1} \circ \phi_\alpha = \mathrm{id}_{U_\alpha}$. So we may define $d\omega$ to be the form whose restriction to U_α is $\phi_\alpha^*(d\omega_\alpha)$, provided this is well-defined on the intersection of coordinate domains:

Definition 5.2.1. For a k-form ω on M, we define $d\omega$ to be the $(k+1)$-form η such that, for each chart (ϕ_α, U_α), $\eta\big|_{U_\alpha} = \phi_\alpha^*(d(\phi_\alpha^{-1})^*(\omega))$.

Proposition 5.2.2. *This is indeed well-defined.*

Proof. If (ϕ_β, U_β) is a second chart then, on $U_\alpha \cap U_\beta$, and its respective images in V_α and V_β, we have

$$\begin{aligned}(\phi_\alpha)^*(d\omega_\alpha) &= (\theta_{\alpha\beta} \circ \phi_\beta)^*(d\omega_\alpha) = (\phi_\beta)^*\left((\theta_{\alpha\beta})^*(d\omega_\alpha)\right)\\ &= (\phi_\beta)^* d\left((\theta_{\alpha\beta})^*(\omega_\alpha)\right),\end{aligned}$$

where we use the fact that, since $\theta_{\alpha\beta}$ is a diffeomorphism between open subsets of $\mathbb{R}^m$, the pull-back $(\theta_{\alpha\beta})^*$ and the exterior derivative commute. Thus

$$\begin{aligned}(\phi_\alpha)^*(d\omega_\alpha) &= (\phi_\beta)^* d\left(((\theta_{\alpha\beta})^* \circ (\phi_\alpha^{-1})^*)(\omega)\right) = (\phi_\beta)^* d\left((\phi_\alpha^{-1} \circ \theta_{\alpha\beta})^*(\omega)\right)\\ &= (\phi_\beta)^* d\left((\phi_\beta^{-1})^*(\omega)\right) = (\phi_\beta)^*(d\omega_\beta).\end{aligned} \qquad \blacksquare$$

Since all definitions are local, local properties remain true globally. Thus $d\omega$, for a form ω on a manifold, will inherit similar properties to those of Proposition 5.1.2. Indeed properties (2), (3) and (4) look identical for forms on a manifold to those on a linear space. Note however that property (1) still only gives $\mathbb{R}$-linearity.

There is an alternative definition of d, which is direct, rather than being obtained by piecing together local definitions and so having to check their independence of a particular coordinate system.

Definition 5.2.3. If ω is a form of degree $k \geqslant 1$ on the manifold M then $d\omega$ is the differential form of degree $k+1$ on M defined by putting $d\omega(X_1, \cdots, X_{k+1})$ equal to

$$\sum_{i=1}^{k+1}(-1)^{i-1}X_i(\omega(X_1,\cdots,\hat{X}_i,\cdots,X_{k+1})) + \sum_{i<j}(-1)^{i+j}\omega([X_i,\, X_j],\, X_1,\cdots,\hat{X}_i,\cdots,\hat{X}_j,\cdots,X_{k+1}),$$

where $\hat{X}_i$ means 'omit X_i'.

Note that, if ω is a 1-form, then the relation $d\omega(X,Y) = X(\omega(Y)) - Y(\omega(X)) - \omega([X,\, Y])$ is immediate from the definition. That it also follows from our earlier definition of the exterior derivative is Exercise 5.5 below. This should give the reader some confidence that the second definition does actually follow from the first. Conversely, starting from the second definition, direct calculation, using the usual global definition of the differential df, namely $df(X) = Xf$, shows that $d(df) = 0$. Furthermore, if $f_1, \cdots, f_k$ and g are smooth functions on M,

$$d(g\, df_1 \wedge \cdots \wedge df_k) = dg \wedge df_1 \wedge \cdots \wedge df_k.$$

This, together with the fact that d is compacible with restriction, that is, $(d\omega)\big|_U = d(\omega\big|_U)$, shows that our first definition does follow from this global one. Hence so too do the remaining formulae from Proposition 5.1.2.

5.3. Manifolds with Boundary.

We now allow our manifolds to have a boundary. The definition is exactly as before except that now we have charts with images which are open subsets of

$$\mathbb{H}^m = \{(x_1, \cdots, x_m) \in \mathbb{R}^m \mid x_1 \leqslant 0\}.$$

We write $\partial\mathbb{H}^m$ for the subset $\{x \in \mathbb{H}^m \mid x_1 = 0\}$ which is naturally isomorphic with $\mathbb{R}^{m-1}$. Then for V open in $\mathbb{H}^m$ we define $f : V \to \mathbb{R}^n$ to be *differentiable* if there is an open set $U \subseteq \mathbb{R}^m$ and a differentiable map $\tilde{f} : U \to \mathbb{R}^n$ such that $V = U \cap \mathbb{H}^m$ and $\tilde{f}|_V = f$. Then for x in V we define the derivative $Df(x)$ to be $D\tilde{f}(x)$.

Proposition 5.3.1. *$Df(x)$ is well-defined.*

Proof. For points not in $\partial\mathbb{H}^m$ this is trivial: no extension is needed. For points in $\partial\mathbb{H}^m$ we may use curves in $\partial\mathbb{H}^m$ to characterise $m-1$ of the basis vectors. For the remaining basis vector we may use a curve normal to the boundary. Then the vector equivalence class of its image is already determined by the image of the part of the curve which lies in $\mathbb{H}^m$. ■

We may similarly define differentiable maps $f : V \to \mathbb{H}^n$, the extension being $\tilde{f} : U \to \mathbb{R}^n$. We then define a (maximal) smooth atlas and hence a smooth *manifold-with-boundary*, and so on, as before.

There is however a new feature:

Definition 5.3.2. A point p of a manifold-with-boundary M is called a *boundary point* if there is a chart (ϕ, U) at p such that $\phi(p)$ is in $\partial\mathbb{H}^m$. The set of all boundary points is called the *boundary* ∂M of M.

Note that manifolds as previously defined, with charts mapping into $\mathbb{R}^m$, are also manifolds-with-boundary: they just have an empty boundary. We may refer to manifolds-without-boundary when we wish to restrict attention to that case.

Proposition 5.3.3. *The boundary ∂M of an m-dimensional manifold-with-boundary is a well-defined manifold (without boundary) of dimension $m-1$ such that the inclusion map is differentiable. An orientation of M induces a unique orientation of ∂M.*

Proof. Consider charts (ϕ_α, U_α) and (ϕ_β, U_β) at p with $\phi_\alpha(p) \notin \partial\mathbb{H}^m$. Thus $\phi_\alpha(p)$ is in $\mathbb{H}^m \setminus \partial\mathbb{H}^m$ which is open in $\mathbb{R}^m$ so we may assume $\phi_\alpha(U_\alpha)$ is open in $\mathbb{R}^m$. Then the coordinate transformation $\theta_{\beta\alpha}$ at $\phi_\alpha(p)$ being differentiable with a differential inverse means that $D\theta_{\beta\alpha}(\phi_\alpha(p))$ is a linear isomorphism. So by the Inverse Function Theorem there is an open neighbourhood of $\phi_\alpha(p)$, open now in $\mathbb{R}^m$, mapped bijectively and diffeomorphically onto an open neighbourhood, open in $\mathbb{R}^m$ still, of $\phi_\beta(p)$. This is also the image of an open neighbourhood of p in U_α and so open in $\mathbb{H}^m$. Thus $\phi_\beta(p)$ is not in $\partial\mathbb{H}^m$, so that

∂M is a well-defined subset of M and for all charts (ϕ_α, U_α) the intersection $V_\alpha = U_\alpha \cap \partial M$ is open in ∂M and $\phi_\alpha|_{V_\alpha}$ maps to $\partial\mathbb{H}^m \cong \mathbb{R}^{m-1}$. Then the restrictions of coordinate transformations on M are coordinate transformations between open subsets of $\mathbb{R}^{m-1}$. Thus ∂M is a differentiable manifold, without boundary, that is having open subsets of $\mathbb{R}^{m-1}$ for its coordinate spaces and, corresponding to an atlas $\mathcal{A}_M$ on M, the boundary ∂M has the atlas

$$\mathcal{A}_{\partial M} = \{(\phi|_{\partial M}, U \cap \partial M) \mid (\phi, U) \in \mathcal{A}_M\}.$$

Using the charts $\phi|_{\partial M}$ on ∂M and ϕ on M at a point p in ∂M, the inclusion of ∂M in M translates into the inclusion of $\partial\mathbb{H}^m$ in $\mathbb{H}^m$ so it is trivially smooth.

We now examine the coordinate transformations from an oriented atlas on M at a point p in ∂M, as illustrated in Figure 5.1.

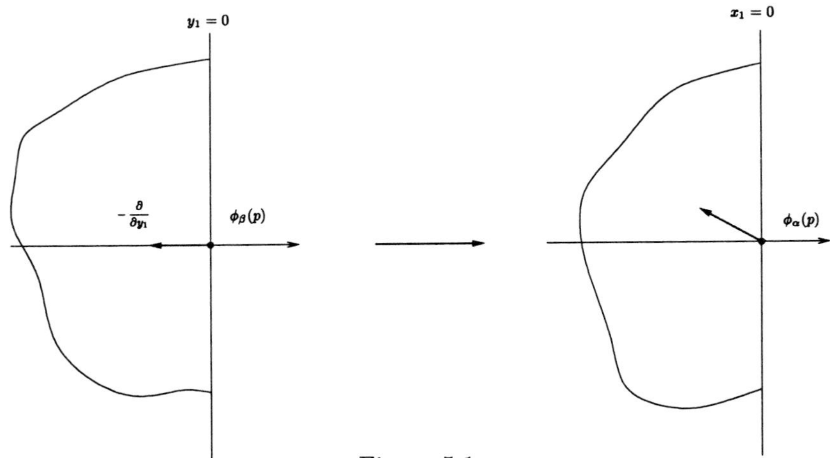

Figure 5.1.

Then $J(\theta_{\alpha\beta})(\phi_\beta(p))$ maps the subspace $\{y_1 = 0\}$ to the subspace $\{x_1 = 0\}$ and also maps the vector $-\frac{\partial}{\partial y_1}$ indicated to a vector with first coordinate strictly negative. These facts are clear from the interpretation of vectors as equivalence classes of curves and the fact that θ_{α_β} maps $\phi_\beta(U_\beta \cap U_\alpha) \cap \partial\mathbb{H}^m$ onto $\phi_\alpha(U_\beta \cap U_\alpha) \cap \partial\mathbb{H}^m$. So, writing $\bar{\theta}_{\alpha\beta}$ for the restriction $\theta_{\alpha\beta}|_{\mathbb{R}^{m-1}}$ of $\theta_{\alpha\beta}$, mapping $\mathbb{R}^{m-1}$ into itself, the Jacobian matrix of the derivative of $\theta_{\alpha\beta}$ has the block decomposition

$$\begin{pmatrix} \frac{\partial x_1}{\partial y_1} & 0 \\ * & J\bar{\theta}_{\alpha\beta} \end{pmatrix}.$$

Then $\det(J(\theta_{\alpha\beta})) > 0$ and $\frac{\partial x_1}{\partial y_1} > 0$ imply that $\det(J(\bar{\theta}_{\alpha\beta})) > 0$ as required for an oriented atlas. ■

For a chart (ϕ, U) on M and its restriction to ∂M we have the commutative Diagram 5.2, where $\iota : \partial M \hookrightarrow M$ is the inclusion and κ is the inclusion $(x_2, \cdots, x_m) \mapsto (0, x_2, \cdots, x_m)$ of $\mathbb{R}^{m-1}$ in $\mathbb{H}^m$.

$$\begin{array}{ccccccc}
\partial M & \supset & W = U \cap \partial M & \overset{\iota}{\hookrightarrow} & U & \subset & M \\
 & & \downarrow \phi|_W & & \downarrow \phi & & \\
 & & \mathbb{R}^{m-1} & \overset{\kappa}{\hookrightarrow} & \mathbb{H}^m & &
\end{array}$$

Diagram 5.2.

Note that the coordinates $\{(x_1, \cdots, x_m) \mid x_1 \leqslant 0\}$, on the coordinate space in $\mathbb{H}^m$ for M, restrict to the coordinates $\{(x_2, \cdots, x_m)\}$ on the corresponding coordinate space for ∂M, when we identify $\mathbb{R}^{m-1}$ with its image in $\mathbb{H}^m$. With that convention we have:

Addendum 5.3.4. $\iota^* \circ \phi^*(dx^i) = (\phi|_W)^*(dx^i)$ *for* $i \geqslant 2$ *and* $\iota^* \circ \phi^*(dx^1) = 0$.

Proof. We have $\kappa_* \left(\frac{\partial}{\partial x_i}\right) = \frac{\partial}{\partial x_i}$ for $i \geqslant 2$. Then

$$(\kappa^*(dx^i))\left(\frac{\partial}{\partial x_j}\right) = dx^i\left(\kappa_*\left(\frac{\partial}{\partial x_j}\right)\right) = \delta^i_j$$

for $j \geqslant 2$. This shows that $\kappa^*(dx^i) = dx^i$ for $i \geqslant 2$ and $\kappa^*(dx^1) = 0$. The result follows by pulling back onto the coordinate neighbourhoods:

$$\begin{aligned}
\iota^* \circ \phi^*(dx^i) &= (\phi|_W)^* \circ \kappa^*(dx^i) = (\phi|_W)^*(dx^i), \qquad \text{for } i \geqslant 2, \\
\iota^* \circ \phi^*(dx^1) &= (\phi|_W)^* \circ \kappa^*(dx^1) = 0.
\end{aligned}$$
■

Note. If we adopt the usual convention of writing dx^i for both of the forms $\phi^*(dx^i)$ and $(\phi|_{\partial M})^*(dx^i)$, these formulae become $\iota^*(dx^i) = dx^i$ for $i \geqslant 2$ and $\iota^*(dx^1) = 0$, as for the corresponding forms on $\mathbb{R}^{m-1}$ and $\mathbb{H}^m$.

5.4. Stokes' Theorem.

We now come to our m-dimensional generalisation of the fundamental theorem of the integral calculus: $\int_a^b f'(x)dx = f(b) - f(a)$.

Theorem 5.4.1. (Stokes) *Let M^m be a compact oriented manifold-with-boundary, let ∂M have the induced orientation, $\iota : \partial M \hookrightarrow M$ be the inclusion and $\omega \in \Omega^{m-1}(M)$ be an $(m-1)$-form on M. Then*

$$\int_{\partial M} \iota^*(\omega) = \int_M d\omega.$$

Proof. Suppose the result valid for a form ω such that the support of ω lies in a coordinate neighbourhood U. Then, for a general $(m-1)$-form ω, let $\{f_k \mid k = 1, \cdots, n\}$ be a partition of unity subordinate to a covering of M by coordinate neighbourhoods $\{U_k\}$ from the oriented atlas on M. Then, for $\omega_k = f_k\,\omega$, we have $\omega = \sum\limits_{k=1}^{n} \omega_k$ and

$$\int_{\partial M} \iota^*(\omega) = \sum_{k=1}^{n} \int_{\partial M} \iota^*(\omega_k) = \sum_{k=1}^{n} \int_M d\omega_k = \int_M d\sum_{k=1}^{n} \omega_k = \int_M d\omega.$$

Thus it remains to prove the special case of the theorem.

So we suppose that $\mathrm{supp}(\omega)$ is contained in the coordinate neighbourhood U for a chart ϕ and that on $V = \phi(U)$

$$(\phi^{-1})^*(\omega) = \sum_{j=1}^{m} a_j dx^1 \wedge \cdots \wedge \widehat{dx^j} \wedge \cdots \wedge dx^m$$

where $\widehat{dx^j}$ indicates that that factor is omitted. Then

$$(\phi^{-1})^*(d\omega) = d\left((\phi^{-1})^*(\omega)\right) = \sum_{j=1}^{m} (-1)^{j-1} \frac{\partial a_j}{\partial x_j}\, dx^1 \wedge \cdots \wedge dx^m$$

and also, writing $W = U \cap \partial M$,

$$(\phi|_W{}^{-1})^*(\iota^*(\omega)) = k^*\left((\phi^{-1})^*(\omega)\right) = (a_1 \circ \iota)\, dx^2 \wedge \cdots \wedge dx^m$$

on $V \cap \partial\mathbb{H}^m$, since $\kappa^*(dx^1) = 0$ and $\kappa^*(dx^i) = dx^i$ for $i \geqslant 2$.

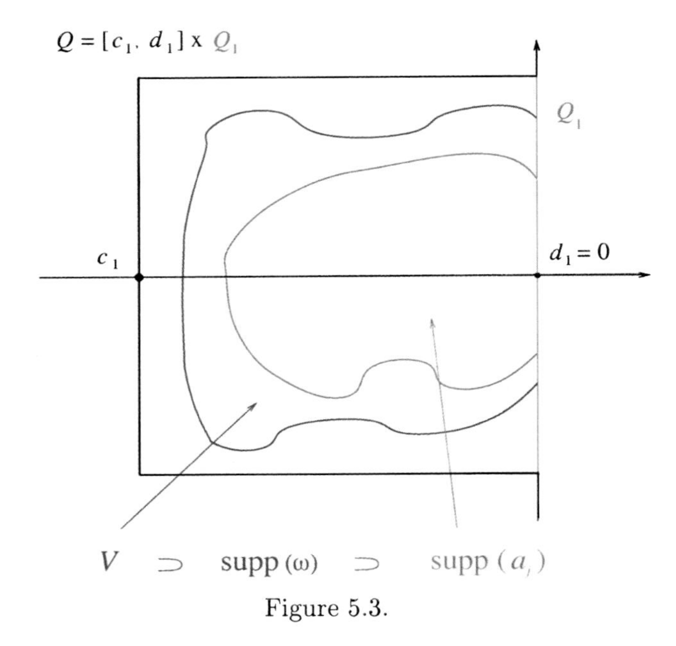

Figure 5.3.

Let V be contained in the product $Q = [c_1, d_1] \times \cdots \times [c_m, d_m]$ where $d_1 = 0$. Extend the functions a_j to Q by defining $a_j(x) = 0$ for $x \notin V$. Then the extension is differentiable on M since a_j is zero in the open set $(\mathrm{supp}(\omega))^c$, the complement of the support of ω, whose union with V is the whole of M. Write Q_j for the product $[c_1, d_1] \times \cdots \times \widehat{[c_j, d_j]} \times \cdots \times [c_m, d_m]$ of all the intervals except the jth, but continue to write x_i for the coordinate in $[c_i, d_i]$ whether it occurs before or after the deleted factor. The case $j = 1$ is illustrated in Figure 5.3. Then

$$\begin{aligned}
\int_M d\omega = \int_U d\omega &= \int_V \sum_{j=1}^{m} (-1)^{j-1} \frac{\partial a_j}{\partial x_j} dx^1 \cdots dx^m \\
&= \sum_{j=1}^{m} \int_Q (-1)^{j-1} \frac{\partial a_j}{\partial x_j} dx^1 \cdots dx^m \\
&= \sum_{j=1}^{m} \int_{Q_j} (-1)^{j-1} \left[a_j(x_1, \cdots, x_{j-1}, d_j, x_{j+1}, \cdots, x_m) \right. \\
&\qquad \left. - a_j(x_1, \cdots, x_{j-1}, c_j, x_{j+1}, \cdots, x_m) \right] dx^1 \cdots \widehat{dx^j} \cdots dx^m,
\end{aligned}$$

by integrating the jth summand with respect to the jth coordinate first, which is possible by Fubini's theorem, see Royden (1968) Chapter 12, Theorem 19. However all these summands except the first vanish on the relevant domain Q_j and $a_1(c_1, x_2, \cdots, x_m)$ is also zero. So this expansion of the integral of $d\omega$ reduces to

$$\begin{aligned}\int_M d\omega &= \int_{Q_1} a_1(0, x_2, \cdots, x_m)\, dx^2 \cdots dx^m \\ &= \int_{V \cap \partial \mathbb{H}^m} (a_1 \circ \iota)(x)\, dx^2 \cdots dx^m = \int_{U \cap \partial M} \iota^*(\omega) \\ &= \int_{\partial M} \iota^*(\omega).\end{aligned}$$ ■

Corollary 5.4.2. *If M^m is a compact orientable manifold without boundary and $\omega \in \Omega^{m-1}(M)$ is an $(m-1)$-form on M, then $\int_M d\omega = 0$.* ■

Proposition 5.4.3. *If M is a compact orientable manifold with boundary, then there is no smooth map $f : M \to \partial M$ such that $f \circ \iota = \mathrm{id}_{\partial M}$.*

Proof. If there were such map f, let ω be a volume $(m-1)$-form on ∂M, arising from the induced orientation of the boundary. Then $d(f^*(\omega)) = f^*(d\omega) = 0$ since $d\omega$ in $\Omega^m(\partial M)$ which is zero. So

$$0 = \int_M d(f^*\omega) = \int_{\partial M} \iota^* \circ f^*(\omega) = \int_{\partial M} (f \circ \iota)^*(\omega) = \int_{\partial M} \omega.$$

But, as ω is a volume form, it will have everywhere positive coefficients in every chart from the orientation atlas on ∂M. Then, if $\{g_i\}$ is a subordinate partition of unity, $\int_{\partial M} g_i\, \omega > 0$ and so

$$\int_{\partial M} \omega = \sum_i \int_{\partial M} g_i\, \omega > 0.$$ ■

Corollary 5.4.4. (Brouwer) *Every differentiable map $g : \bar{B} \to \bar{B}$ of the closed unit ball in $\mathbb{R}^m$ into itself has a fixed point.*

Proof. Otherwise each directed line $\overrightarrow{g(p)p}$ intersects the boundary ∂B in a unique point $f(p)$ which, for p in ∂B, is p itself. Thus f, assuming it were smooth, would satisfy the hypotheses of Proposition 5.4.3 and so cannot exist. Hence neither can g exist without fixed points.

It remains to show that f is indeed smooth. So write $f(x) = x + \lambda u$ where u is the unit vector $\frac{x - g(x)}{\|x - g(x)\|}$ and $\lambda > 0$ is chosen so that $\| f(x) \| = 1$, that is, $x \cdot x + 2\lambda(x \cdot u) + \lambda^2 = 1$, from which

$$\lambda = -(x \cdot u) + \left((x \cdot u)^2 + 1 - x \cdot x\right)^{1/2}.$$

Since $x \neq g(x)$, all is smooth unless $x \cdot u = 0$ and $x \cdot x = 1$. But that would require $x \in \mathbf{S}^{m-1}$ and $(x - g(x)) \cdot x = 0$, which would force $g(x)$ to lie outside B. ∎

Remark 5.4.5. Both Proposition 5.4.3 and Corollary 5.4.4 are valid for continuous maps with an analogous proof using, say, singular homology. As we shall see later, integrals of forms over submanifolds can be interpreted as homology elements.

5.5. Bubbling Forms.

In this section we produce a useful result concerning top dimensional forms on a manifold. This says that, for any m-form ω on an m-dimensional manifold and any given open subset U of M, there is an $(m-1)$-form η such that $\omega + d\eta$ has support on U. This will be used for example to identify the top-dimensional cohomology space in the next chapter. The pre-cursor of this result is Theorem 5.5.5 below.

As usual we first prove this result locally on $\mathbb{R}^n$ and then globalise it using a partition of unity on M. Thus the first lemma we require involves an elementary manipulation of multiple integrals on Euclidean spaces. We denote by $\mathbf{I}^m$ the unit hypercube $\{x \mid 0 \leqslant x_i \leqslant 1,\ i = 1, \cdots, m\} \subset \mathbb{R}^m$.

Lemma 5.5.1. *If f is a smooth function on $\mathbb{R}^m$ with $\operatorname{supp}(f) \subseteq \mathbf{I}^m$ and $\int_{\mathbb{R}^m} f(x)\, dx^1 \cdots dx^m = 0$, then there exist functions $f_i \in C^\infty(\mathbb{R}^m)$ such that $\operatorname{supp}(f_i) \subseteq \mathbf{I}^m$ and $\sum_{j=1}^{m} \frac{\partial f_j}{\partial x_j} = f$.*

Proof. Induction on m. For the case that $m = 1$, we have f such that $\int_{-\infty}^{\infty} f(t)dt = 0$ and so may choose $f_1(x) = \int_{-\infty}^{x} f(t)dt$. Then, assuming the result for $\mathbb{R}^{m-1}$, we define

$$g(x_1, \cdots, x_{m-1}) = \int_{-\infty}^{\infty} f(x_1, \cdots, x_{m-1}, t)\, dt.$$

By differentiating under the integral sign we see that g is in $C^\infty(\mathbb{R}^{m-1})$ and $\operatorname{supp}(g) \subseteq \mathbf{I}^{m-1}$ while, by Fubini's Theorem,

$$\int_{\mathbb{R}^{m-1}} g\, dx^1 \cdots dx^{m-1} = \int_{\mathbb{R}^m} f\, dx^1 \cdots dx^m = 0.$$

So by induction there exist $g_1, \cdots, g_{m-1}$ with $\text{supp}(g_i) \subseteq \mathbf{I}^{m-1}$ and such that $\sum_{j=1}^{m-1} \frac{\partial g_j}{\partial x_j} = g$. Let $f_j(x_1, \cdots, x_{m-1}, x_m) = g_j(x_1, \cdots, x_{m-1})\rho(x_m)$, where $\text{supp}(\rho) \subseteq \mathbf{I}^1$ and $\int_{-\infty}^{\infty} \rho(t)\, dt = 1$, and let $h = f - \sum_{j=1}^{m-1} \frac{\partial f_j}{\partial x_j}$. Then

$$f_m(x_1, \cdots, x_{m-1}, x_m) = \int_{-\infty}^{x_m} h(x_1, \cdots, x_{m-1}, t)dt$$

has $\frac{\partial f_m}{\partial x_m} = h$ so that $f = \sum_{j=1}^{m} \frac{\partial f_j}{\partial x_j}$. Then f_m is in $\mathcal{C}^\infty(\mathbb{R}^m)$ and to see that $\text{supp}(f_m) \subseteq \mathbf{I}^m$ we need $\int_{-\infty}^{\infty} h(x_1, \cdots, x_{m-1}, t)dt = 0$. But, for $j < m$, $\frac{\partial f_j}{\partial x_j} = \frac{\partial g_j}{\partial x_j}$ so

$$\begin{aligned} h(x_1, \cdots, x_{m-1}, t) &= f(x_1, \cdots, x_{m-1}, t) - \sum_{j=1}^{m-1} \frac{\partial g_j}{\partial x_j}(x_1, \cdots, x_{m-1}, t)\rho(t) \\ &= f(x_1, \cdots, x_{m-1}, t) - g(x_1, \cdots, x_{m-1})\rho(t) \end{aligned}$$

and so

$$\begin{aligned} &\int_{-\infty}^{\infty} h(x_1, \cdots, x_{m-1}, t)dt \\ &\quad = \int_{-\infty}^{\infty} f(x_1, \cdots, x_{m-1}, t)dt - g(x_1, \cdots, x_{m-1}) \int_{-\infty}^{\infty} \rho(t)\, dt = 0 \end{aligned}$$

as required. ■

Corollary 5.5.2. *Let $\omega \in \Omega^m(\mathbb{R}^m)$ be such that* $\text{supp}(\omega) \subseteq \mathbf{I}^m$ *and* $\int_{\mathbb{R}^m} \omega = 0$. *Then we may find $\eta \in \Omega^{m-1}(\mathbb{R}^m)$ such that $d\eta = \omega$ and* $\text{supp}(\eta) \subseteq \mathbf{I}^m$ *also.*

Proof. If

$$\eta = \sum_{j=1}^{m} (-1)^{j-1} f_j(x)\, dx^1 \wedge \cdots \wedge \hat{dx^j} \wedge \cdots \wedge dx^m$$

and $\omega = f(x)dx^1 \wedge \cdots \wedge dx^m$, then $\omega = d\eta$ if and only if $f = \sum_{j=1}^{m} \frac{\partial f_j}{\partial x_j}$. So, given f, the existence of such f_j with $\text{supp}(f_j) \subseteq \mathbf{I}^m$ is granted by Lemma 5.5.1. ■

Our second lemma is a preliminary result for manifolds. It uses forms in the image of d to transfer the support of an m-form from one coordinate neighbourhood to another, along a chain of over-lapping coordinate neighbourhoods. It will be convenient to work with surjective charts $\phi_\alpha : V_\alpha \to \mathbb{R}^m$. We may produce such a chart at each point p of a manifold as follows: for any chart ϕ on a coordinate neighbourhood containing p first translate the image so that $\phi(p) = 0$; since the coordinate space is open, it contains some ball B of radius r say; then $\phi|_{\phi^{-1}(B)}$ followed by a diffeomorphism of B onto $\mathbb{R}^m$ is a surjective chart at p. Our functions and forms will generally be supported on the 'central domains' $U = \phi^{-1}(\mathbf{I}^m)$ of such charts. Throughout the remainder of this section all coordinate neighbourhoods V_α will be assumed to have such surjective charts ϕ_α and central domains U_α.

Lemma 5.5.3. *Let M^m be connected and V_0, V_∞ be two coordinate neighbourhoods and $\omega_0 \in \Omega^m(M)$ be an m-form on M such that* $\operatorname{supp}(\omega_0) \subseteq U_0$. *Then there is an $(m-1)$-form η on M such that* $\operatorname{supp}(\omega_0 + d\eta) \subseteq U_\infty$.

Proof. Let (V_j, U_j), $j = 0, 1, \cdots, r$, be a sequence of such coordinate neighbourhoods and their central domains with $U_{j-1} \cap U_j \neq \emptyset$ for $j = 1, \cdots, r$ and $V_r = V_\infty$. Such a sequence exists by the connectivity of M.

We then define, sequentially, m-forms ω_0, ω_1, $\cdots$, ω_r, starting with the given ω_0, with ω_j having support contained in U_j, and $(m-1)$-forms η_1, $\cdots, \eta_r$, which may also have supports contained in $U_1, \cdots, U_r$ respectively though we do not need that, such that $d\eta_j = \omega_j - \omega_{j-1}$. In fact choosing *any* m-form ω_j, $j = 1, 2, \cdots, r$, with support contained in $U_j \cap U_{j-1}$ and such that $\int_{U_{j-1}} \omega_j = \int_{U_{j-1}} \omega_{j-1}$, which is possible by re-scaling ω_j, we have

$$\int_{V_{j-1}} (\omega_j - \omega_{j-1}) = 0 \text{ and } \operatorname{supp}(\omega_j - \omega_{j-1}) \subseteq U_{j-1}.$$

Transforming to the coordinate space, using Corollary 5.5.2 and pulling back using the coordinate chart gives η_j.

Then, writing $\eta = \sum_{j=1}^{r} \eta_j$, we have $d\eta = \omega_r - \omega_0$ and so $\operatorname{supp}(\omega_0 + d\eta) = \operatorname{supp}(\omega_r) \subseteq U_r = U_\infty$ as required. ■

Proposition 5.5.4. *Let M^m be a compact connected manifold and $\phi : V \to \mathbb{R}^m$ a surjective chart with $U = \phi^{-1}(\mathbf{I}^m)$. Then for any m-form $\omega \in \Omega^m(M)$, there is an $(m-1)$-form η such that* $\operatorname{supp}(\omega + d\eta)$ *is contained in U.*

Proof. Since M is compact it may be covered by finitely many central domains of surjective charts, and we can use a subordinate partition of unity to write ω

as a finite sum of forms ω_i supported on these central domains. Then, applying Lemma 5.5.3 to each of the ω_i, let η_i be such that $\mathrm{supp}(\omega_i + d\eta_i) \subset U$. Then, letting $\eta = \sum_i \eta_i$, we have

$$\mathrm{supp}(\omega + d\eta) = \mathrm{supp}\left(\sum_i \omega_i + d\eta_i\right) \subset \bigcup_i \mathrm{supp}(\omega_i + d\eta_i) \subset U. \qquad \blacksquare$$

Finally we get our result.

Theorem 5.5.5. *If M is a compact connected oriented manifold without boundary of dimension m and $\omega \in \Omega^m(M)$ such that $\int_M \omega = 0$, then there is an $(m-1)$-form η such that $d\eta = \omega$.*

Proof. By Proposition 5.5.4, there is an $\eta_1 \in \Omega^{m-1}(M)$ such that $\mathrm{supp}(\omega + d\eta_1) \subset U$, for the central domain U of some surjective chart. But $\int_U(\omega + d\eta_1) = \int_M \omega + \int_M d\eta_1 = 0$ by hypothesis and (a corollary of) Stokes' Theorem respectively. So, by Corollary 5.5.2 applied on the coordinate space and then pulled back to the coordinate neighbourhood, there is an $(m-1)$-form η_2 such that $d\eta_2 = \omega + d\eta_1$. So $\omega = d(\eta_2 - \eta_1)$. $\blacksquare$

Corollary 5.5.6. *If M is a compact connected oriented manifold without boundary of dimension m, then there is a surjective linear map*

$$I : \Omega^m(M^m) \longrightarrow \mathbb{R}$$

whose kernel is $d(\Omega^{m-1}(M^m))$, the image under the exterior derivative of the space of $(m-1)$-forms on the manifold.

Proof. For an m-form ω, we define $I(\omega) = \int_M \omega$. The theorem asserts that, if $I(\omega) = 0$, then ω is in the image of the exterior derivative. Conversely $\omega_M = d\eta$ implies that $\int_M \omega_M = \int_M d\eta = \int_{\partial M} \iota^*(\eta) = 0$, since $\partial M = \emptyset$. To see that I is surjective we recall, from the proof of Proposition 5.4.3, that the volume form ω_M determined by the orientation has non-zero integral over M. So we may simply rescale it so that $\int_M \omega_M = 1$ or, for that matter, any other value $\lambda \in \mathbb{R}$. $\blacksquare$

In the next chapter we shall obtain an important re-interpretation of this corollary.

5.6. Exercises for Chapter 5.

5.1. We may use the standard inner product on $\mathbb{R}^n$ to define an isomorphism between $\mathbb{R}^n$ and its dual and hence a 1-1 correspondence between vector fields and 1-forms, where the vector field X on $U \subset \mathbb{R}^m$ corresponds to the 1-form $\omega = \vartheta(X)$ defined by

$$\omega_p(Y) = \langle X(p), Y \rangle, \qquad \text{for each } Y \in \tau_p(\mathbb{R}^m).$$

(*i*) Show that, if $f : U \to \mathbb{R}$, then the vector field $\vartheta^{-1}(df)$ is

$$\operatorname{grad} f = \frac{\partial f}{\partial x_i} e_i$$

where e_i is the standard basis of $\mathbb{R}^m$.

(*ii*) If $X(x)$ and $Y(x)$ are the vector fields $a^i(x)\, e_i$ and $b^i(x)\, e_i$ on $U \subset \mathbb{R}^3$, calculate $\vartheta^{-1} * d\vartheta(X)$ and $\vartheta^{-1} * (\vartheta(X) \wedge \vartheta(Y))$, where $*$ is defined in Exercise 4.5.

5.2. Let

$$\omega = a(x, y, z)\, dx + b(x, y, z)\, dy + c(x, y, z)\, dz$$

be a 1-form on $\mathbb{R}^3$ such that $d\omega = 0$. Show that $\omega = df$ where

$$f(x, y, z) = \int_0^1 \{x\, a(tx, ty, tz) + y\, b(tx, ty, tz) + z\, c(tx, ty, tz)\}\, dt.$$

[*Hint:* $a(x, y, z) = \int_0^1 \frac{d}{dt}\{t\, a(tx, ty, tz)\}\, dt$.]

5.3. Let M be a compact 3-dimensional submanifold-with-boundary of $\mathbb{R}^3$, $f : M \hookrightarrow \mathbb{R}^3$ be the inclusion and

$$\omega = \frac{1}{3}\{x\, dy \wedge dz + y\, dz \wedge dx + z\, dx \wedge dy\}.$$

Show that $d(\omega/r^3) = 0$ on $\mathbb{R}^3 \setminus \{0\}$, where $r^2 = x^2 + y^2 + z^2$.
Show also that

$$\int_{\partial \mathbf{M}} f^*(\omega) = \operatorname{vol}(M),$$

and deduce that there is a 2-form η on $\mathbf{S}^2$ such that $d\eta = 0$ and but $\eta \neq d\phi$ for any 1-form ϕ.

5.4. Let X be a vector field on a manifold M and $f \in \mathcal{C}^\infty(M)$. Define $Xf \in \mathcal{C}^\infty(M)$ by

$$(Xf)(p) = f_*(X(p)) = df(X(p)).$$

Show that X thus acts linearly on $\mathcal{C}^\infty(M)$, and that

$$X(fg) = g\,(Xf) + f\,(Xg),$$

that is, X determines a derivation of $\mathcal{C}^\infty(M)$. Show conversely that any such derivation determines a vector field. Deduce that

$$[X,Y]f = X(Yf) - Y(Xf)$$

determines a vector field $[X,Y]$ on M.

5.5. Using Definition 5.2.1 or, equivalently, the local version 5.1.1, prove that, for a 1-form ω and vector fields X, Y on M,

$$d\omega(X,Y) = X(\omega(Y)) - Y(\omega(X)) - \omega([X,Y]).$$

CHAPTER 6

DE RHAM COHOMOLOGY

The basic, indeed defining, property of the exterior derivative is that $d^2 = 0$. This means that the sequence of spaces of forms on a manifold and exterior derivative maps between them form what is known in algebraic topology as a cochain complex. This leads immediately to the de Rham cohomology spaces which are therefore invariants of the manifold and which will help us to distinguish between, and prove results about, manifolds in general. These spaces are non-zero only in the degrees from zero up to the dimension of the manifold and, with only limited conditions on the manifold, we are able to calculate them in the lowest and highest degrees, the latter being the promised re-interpretation of the final result in Chapter 5. However to compute the remaining de Rham cohomology requires more effort.

Hence in section 2 we recall how cochain maps between cochain complexes induce linear maps between the corresponding cohomology spaces and how cochain homotopies between cochain maps imply that they induce the same linear maps. We can relate the former result to manifolds by observing that the pull-back homomorphism on forms induced by a map between manifolds is indeed a cochain map. Cochain homotopies will arise from homotopies between maps. However that is a major result and we defer the proof until the next section, here merely recalling the relevant definitions. The main ingredient in the proof is a family of linear maps from k-forms on $M \times I$ to $(k-1)$-forms on M that, when composed with the pull-back of a homotopy, gives rise to a cochain homotopy.

This result, generally known as the Poincaré Lemma although that strictly only refers to a special case, has the important consequence that homotopy equivalent manifolds have isomorphic de Rham cohomology. While this, in some real sense, reduces the number of manifolds for which we have to calculate the cohomology, it does not actually take us far in those calculations. To that end, in section 4, we establish the Mayer-Vietoris exact sequence that expresses the cohomology of a manifold in terms of that of two open submanifolds whose union is the whole manifold. To illustrate that this is an effective tool we use it to calculate all the cohomology spaces of spheres and tori in the next two sections.

Classically, and historically, cohomology theories arose as the duals of homology theories. In the final section of this chapter we reverse that procedure and show how to obtain the dual de Rham homology theory.

6.1. Basic Definitions.

Writing, temporarily, d_k for the exterior derivative on the real vector space $\Omega^k(M)$ of k-forms, overlooking for present purposes its extra structure as a $\mathcal{C}^\infty(M)$-module, we have the sequence of vector spaces and linear maps

$$\cdots \longrightarrow \Omega^{k-1}(M) \xrightarrow{d_{k-1}} \Omega^k(M) \xrightarrow{d_k} \Omega^{k+1}(M) \longrightarrow \cdots .$$

The fact that $d_k \circ d_{k-1} = 0$ for all k makes this into what is known as a *cochain complex.* This condition is equivalent to the statement that the image of d_{k-1} is a subspace of the kernel of d_k. Forms in the kernel of d are called *closed* and those in the image called *exact.* Then the de Rham cohomology 'groups' are the quotients of the spaces of closed forms by the spaces of exact forms.

Definition 6.1.1. The *kth de Rham cohomology group* or *space* of the manifold M is the quotient space

$$H^k_{\mathrm{dR}}(M) = \frac{\ker(d_k)}{\mathrm{im}(d_{k-1})} = \frac{\{\text{closed } k\text{-forms}\}}{\{\text{exact } k\text{-forms}\}}.$$

We write $[\omega]$ for the element of $H^k_{\mathrm{dR}}(M)$ represented by the closed k-form ω, and refer to it as the *cohomology class of* ω.

In general we do not require M to be connected, compact or orientable. However, for many specific results, we shall require one or more of these properties. When we require all three we shall abbreviate them to 'c.c.o.' Since de Rham cohomology is the only cohomology theory that we shall consider until the end of Chapter 7, we shall usually drop the suffix.

Note that $H^k(M) = 0$ if and only if every closed k-form is exact. Thus we may think of the cohomology space as the 'obstruction to exactness' for closed forms.

Although we shall not be making any use of it in this book, it is an important fact that the product structure on the spaces of forms is inherited by the cohomology spaces. For closed k-forms ω_i, we note that $[\omega_1 \wedge \omega_2]$ depends only on the classes $[\omega_1]$ and $[\omega_2]$, since $(\omega_1 + d\eta_1) \wedge (\omega_2 + d\eta_2)$ is equal to $\omega_1 \wedge \omega_2 + d(\eta_1 \wedge \omega_2 + (-1)^{\deg(\omega_1)}\omega_1 \wedge \eta_2 + \eta_1 \wedge d\eta_2)$ since $d\omega_1 = d\omega_2 = 0$ by assumption. Thus we may define $[\omega_1] \wedge [\omega_2]$ to be $[\omega_1 \wedge \omega_2]$, making the direct sum

$$H^*(M) = \bigoplus_{k=0}^{m} H^k(M)$$

into a graded commutative algebra, since the degree of $[\omega_1] \wedge [\omega_2]$ is equal to $\deg([\omega_1]) + \deg([\omega_2])$ and

$$[\omega_1] \wedge [\omega_2] = (-1)^{\deg(\omega_1)\deg(\omega_1)}[\omega_2] \wedge [\omega_1]$$

on account of the similar relation between the representative forms $\omega_1 \wedge \omega_2$ and $\omega_2 \wedge \omega_1$.

Before entering on less trivial calculations we note two important cases for which we can already compute the cohomology. The first is the 0th cohomology space.

Proposition 6.1.2. *If M is connected, then $H^0(M) = \mathbb{R}$.*

Proof. Since there are no forms of degree -1, no 0-form can be exact. Thus the 0th cohomology space is isomorphic with the space of closed 0-forms. However a 0-form, that is a function f, is closed if and only if $df = 0$. That is, if and only if, in every coordinate space, all its partial derivatives vanish and, that is, if and only if it is everywhere locally constant. Since M is connected, that is if and only if $f(p) = c \in \mathbb{R}$ for all p on M. ■

More generally, if M is not connected, there will be one direct summand of $H^0(M)$ isomorphic with $\mathbb{R}$ for each connected component of M.

The second result that we can obtain without the need for further machinery is the calculation of the cohomology space of degree equal to the dimension of the manifold. Since, for an m-manifold there are forms of degree m and none of degree greater than m that is the highest degree in which the cohomology may be non-zero. Frequently $H^m(M^m)$ will be zero, but there is an important special case in which it will be non-zero. The following is the promised re-interpretation of Corollary 5.5.6.

Theorem 6.1.3. *If M is a compact connected oriented manifold without boundary of dimension m, then $H^m(M^m) = \mathbb{R}$.*

Proof. We only need to observe that any form of degree equal to the dimension m of M is necessarily closed since there are no non-zero $(m+1)$-forms. ■

For such a c.c.o. manifold without boundary and any m-form, not necessarily a volume form, such that $\int_M \omega_M = 1$, the class $[\omega_M] \in H^m(M^m)$ is called the *fundamental cohomology class* of M. Then, if instead $\int_M \omega = \lambda$, it follows that $\int_M (\omega - \lambda\omega_M) = 0$. So $\omega - \lambda\omega_M = d\eta$ for some $(m-1)$-form η and then $[\omega] = [\lambda\omega_M] = \lambda[\omega_M]$. Thus there is a cohomology class for each real $\lambda \in \mathbb{R}$ and the class $[\omega]$ of ω is determined by its integral $\int_M \omega$ over M.

Note that, translated into terms of cohomology, Proposition 5.5.4 says, in effect, that any top dimensional cohomology class of a compact connected manifold, whether orientable or not, may be represented by a form supported in a given coordinate domain. This will be useful in future computations.

6.2. Cochain Maps and Cochain Homotopies.

In this section we recall the main results from algebraic topology concerning cochain complexes. Although they are valid for general cochain complexes, we shall just state them for the context that concerns us here.

Definition 6.2.1. Given two cochain complexes $\Omega^*(M)$ and $\Omega^*(N)$, a ladder of linear maps f^* between them forming *commutative squares* as in Diagram 6.1 is called a *cochain map*. By the 'commutativity of the squares' we mean that $f^* \circ d_k = d_k \circ f^*$ for all k.

$$\begin{array}{ccccccccc}
\cdots & \longrightarrow & \Omega^{k-1}(M) & \xrightarrow{d_{k-1}} & \Omega^k(M) & \xrightarrow{d_k} & \Omega^{k+1}(M) & \longrightarrow & \cdots \\
 & & \downarrow{\scriptstyle f^*} & & \downarrow{\scriptstyle f^*} & & \downarrow{\scriptstyle f^*} & & \\
\cdots & \longrightarrow & \Omega^{k-1}(N) & \xrightarrow{d_{k-1}} & \Omega^k(N) & \xrightarrow{d_k} & \Omega^{k+1}(N) & \longrightarrow & \cdots
\end{array}.$$

Diagram 6.1.

Proposition 6.2.2. *If $f : N \to M$ is a smooth map of smooth manifolds, with or without boundary, then the pull-backs f^* form a cochain map between their cochain complexes of spaces of forms.*

Proof. This is precisely what we have illustrated in Diagram 6.1. ■

The importance of cochain maps is clear from the following result, again stated just for our specific case.

Corollary 6.2.3. *If $f : N \to M$ is a smooth map and ω a closed k-form on M, then $f^*(\omega)$ is closed and the class $[f^*(\omega)] \in H^k(N)$ depends only on the class $[\omega]$.*

Proof. Note first that, since $d(f^*(\omega)) = f^*(d(\omega)) = 0$, $f^*(\omega)$ is indeed closed and so defines a class $[f^*(\omega)]$. Then, if the class $[\omega]$ is represented instead by $\omega + d\eta$ for some $(k-1)$-form $\eta \in \Omega^{k-1}(M)$, we have

$$f^*(\omega + d\eta) = f^*(\omega) + f^*(d\eta) = f^*(\omega) + d(f^*(\eta)).$$

So $[f^*(\omega + d\eta)] = [f^*(\omega)]$ as required. ■

What we see from this corollary is that a cochain map $f^* : \Omega^*(M) \to \Omega^*(N)$ induces maps

$$(f^*)_* : H^k(M) \longrightarrow H^k(N); \quad [\omega] \mapsto [f^*(\omega)],$$

which are linear since the pull-backs f^* are linear. Usually, when no ambiguity can arise, we shall write f_* rather than the full $(f^*)_*$. The lower '$*$' is chosen to reflect the fact that $(f^*)_*$ goes in the same direction as f^*. It is then an automatic consequence of the manner of definition of these maps that they form a contravariant functor:

Proposition 6.2.4. *If $f^* : \Omega^*(M) \to \Omega^*(N)$ and $g^* : \Omega^*(N) \to \Omega^*(P)$ are cochain maps then $(g^* \circ f^*)_* = (g^*)_* \circ (f^*)_*$. Also $(\mathrm{id}^*)_* = \mathrm{id}$.* ■

Definition 6.2.5. If $f^*, g^* : \Omega^*(M) \to \Omega^*(N)$ are two cochain maps, a map $h : \Omega^*(M) \to \Omega^*(N)$ such that $dh + hd = f^* - g^*$ is called a *cochain homotopy between f and g.*

Note that, since f^* and g^* preserve degree and d increases it by one, this implies that h must be of degree -1, that is, $h : \Omega^k(M) \to \Omega^{k-1}(N)$.

Proposition 6.2.6. *If $f^*, g^* : \Omega^*(M) \to \Omega^*(N)$ are cochain homotopic cochain maps, then $f_* = g_* : H^k(M) \to H^k(N)$.*

Proof. We have Diagram 6.2, in which the chain homotopy maps are those pointing downward and to the left. Let $\omega \in \Omega^k(M)$ be closed and h be the chain homotopy between f^* and g^*. Then

$$f^*(\omega) - g^*(\omega) = d_{k-1}h(\omega) + h(d_k\omega) = d_{k-1}(h(\omega)).$$

So $f_*[\omega] = [f^*(\omega)] = [g^*(\omega)] = g_*[\omega]$. ■

$$\begin{array}{ccccccccc}
\cdots & \longrightarrow & \Omega^{k-1}(M) & \xrightarrow{d_{k-1}} & \Omega^k(M) & \xrightarrow{d_k} & \Omega^{k+1}(M) & \longrightarrow & \cdots \\
 & & f^*\downarrow\downarrow g^* & \swarrow & f^*\downarrow\downarrow g^* & \swarrow & f^*\downarrow\downarrow g^* & & \\
\cdots & \longrightarrow & \Omega^{k-1}(N) & \xrightarrow{d_{k-1}} & \Omega^k(N) & \xrightarrow{d_k} & \Omega^{k+1}(N) & \longrightarrow & \cdots
\end{array}$$

Diagram 6.2.

Just as a differentiable map between manifolds leads to pull-backs that together form a cochain map, in the next section we shall show that homotopies between maps produce cochain homotopies between such cochain maps. In preparation for that we recall the relevant definitions. We shall as usual assume that the maps involved are smooth, although they are generally only required to be continuous. For maps between differentiable manifolds this is no loss of generality: continuous maps may always be approximated, to any given reasonable measure of accuracy, by smooth ones – we leave this as an exercise at the end of this chapter.

We denote by ι_t the inclusion

$$\iota_t : N \longrightarrow N \times \mathbb{R}; \quad x \mapsto (x, t).$$

Definition 6.2.7. A map $h : N \times \mathbb{R} \to M$ is a *homotopy between $f_0 = h \circ \iota_0$ and $f_1 = h \circ \iota_1$*. When such a homotopy exists we write $f_0 \simeq f_1$. Manifolds N and M are *homotopy equivalent* if there exist maps $f : N \to M$ and $g : M \to N$ such that $f \circ g \simeq \mathrm{id}_M$ and $g \circ f \simeq \mathrm{id}_N$. Then the maps f and g are called *inverse homotopy equivalences.* A manifold M is *contractible* if $\mathrm{id}_M \simeq c$ where c is any constant map of M to one of its points.

Examples 6.2.8. 1) $\iota_0 : M \to M \times \{0\} \subseteq M \times \mathbb{R}^n$ and $\pi : M \times \mathbb{R}^n \to M$; $(p, v) \mapsto p$ are inverse homotopy equivalences: $\pi \circ \iota_0 = \mathrm{id}_M$ so, trivially, no homotopy is needed. While

$$h : M \times \mathbb{R}^n \times \mathbb{R} \longrightarrow M \times \mathbb{R}^n; \quad (p, v, t) \mapsto (p, tv)$$

is a homotopy from $\iota_0 \circ \pi : (p, v) \mapsto (p, 0)$ to $\mathrm{id}_{M \times \mathbb{R}^n}$.

2) $\mathbb{R}^n$ is contractible by the homotopy

$$h(p, t) = p + t(p_0 - p).$$

3) A manifold M is contractible if and only if it is homotopy equivalent to a point. In fact, if $f : \{*\} \to M$ is *any* map from the one-point space $\{*\}$ into M with $f(*) = p$, then $\{*\} \xrightarrow{f} M \xrightarrow{g} \{*\}$ is necessarily the identity on the point $*$ and $M \xrightarrow{g} \{*\} \xrightarrow{f} M$ is the constant map $c : M \to p$. So $f \circ g \simeq \mathrm{id}_M$, which is now equivalent to $M \simeq \{*\}$, if and only if $c \simeq \mathrm{id}_M$, which is the definition of M being contractible.

6.3. The Poincaré Lemma.

What we shall now show, by obtaining chain homotopies from homotopies between maps, is that homotopy equivalent manifolds have isomorphic de Rham cohomology groups in every degree. As a special case, all the de Rham groups of a connected contractible manifold M vanish except for $H^0(M) \cong \mathbb{R}$, and as a special case of that we have the classical Poincaré Lemma: any closed k-form on $\mathbb{R}^m$ for $k > 0$ is exact.

We need two preliminary lemmas.

Lemma 6.3.1. *Writing $\pi : M \times \mathbb{R} \to M$ for the projection $(p, t) \mapsto p$. A k-form ω on $M \times \mathbb{R}$ may be written uniquely as*

$$\omega = \omega_1 + dt \wedge \omega_2$$

where $\omega_1(v_1, \cdots, v_k) = 0$, if $\pi_*(v_j) = 0$ for some j, and ω_2 has the similar property.

Proof. Locally, for coordinates $(x_1, \cdots, x_m, t)$, $\pi_*(v) = 0$ if and only if $v = \lambda\frac{\partial}{\partial t}$, so the specification of the ω_i becomes $\omega_i(v_1, \cdots, v_k) = 0$ whenever, for some j, $v_j = \lambda\frac{\partial}{\partial t}$. The local basis of k-forms comprises terms $dx^I = dx^{i_1} \wedge \cdots \wedge dx^{i_k}$

and $dt \wedge dx^J$, where $dx^J = dx^{j_1} \wedge \cdots \wedge dx^{j_{k-1}}$. However $dx^i\left(\frac{\partial}{\partial t}\right) = 0$ for all i and $dt\left(\frac{\partial}{\partial t}\right) = 1$. Thus no term $dt \wedge dx^J$ satisfies the specification, whereas all the terms dx^I do. So

$$\omega = \sum_I a_I dx^I + dt \wedge \sum_J b_J dx^J = \omega_1 + dt \wedge \omega_2$$

is the required decomposition, and the only possible one. The definition extends to M by defining it on all coordinate neighbourhoods. It is globally well-defined because of its invariant characterisation and unique since it is locally unique. ■

We now use this decomposition to define

$$I : \Omega^k(M \times \mathbb{R}) \longrightarrow \Omega^{k-1}(M)$$

by

$$I(\omega)(p)(v_1, \cdots, v_{k-1}) = \int_0^1 \{\omega_2(p,t)((\iota_t)_*(v_1), \cdots, (\iota_t)_*(v_{k-1}))\}\, dt.$$

This is an operation of frequent use in differential geometry known as 'integration along the fibres', the fibres here being the various copies $\{p\} \times \mathbb{R}$ of $\mathbb{R}$. In our case we only integrate along the segment $\{p\} \times [0,\, 1]$ of the fibre. This provides a chain homotopy between ι_0^* and ι_1^* which will be the foundation of all our chain homotopies:

Lemma 6.3.2. $\iota_1^*(\omega) - \iota_0^*(\omega) = d(I(\omega)) + I(d\omega)$.

Proof. We use local coordinates $(x_1, \cdots, x_m)$ on M and $(x_1, \cdots, x_m, t)$ on $M \times \mathbb{R}$, hoping that the use of different coordinates with the same name will not be confusing. Since $(\iota_t)_*\left(\frac{\partial}{\partial x_i}\right) = \frac{\partial}{\partial x_i}$,

$$\iota_t^*(dt)\left(\frac{\partial}{\partial x_i}\right) = dt\left((\iota_t)_*\left(\frac{\partial}{\partial x_i}\right)\right) = dt\left(\frac{\partial}{\partial x_i}\right) = 0$$

for all i and

$$\iota_t^*(dx^i)\left(\frac{\partial}{\partial x_j}\right) = dx^i\left((\iota_t)_*\left(\frac{\partial}{\partial x_j}\right)\right) = dx^i\left(\frac{\partial}{\partial x_j}\right) = \delta_j^i,$$

from which it follows that $\iota_t^*(dt) = 0$ and $\iota_t^*(dx^i) = dx^i$. Then, since the pull back is an algebra homomorphism,

$$\iota_t^*(dx^{i_1} \wedge \cdots \wedge dx^{i_k}) = dx^{i_1} \wedge \cdots \wedge dx^{i_k}.$$

Since I is certainly $\mathbb{R}$-linear it suffices to consider the two cases $\omega = \omega_1$ and $\omega = dt \wedge \omega_2$ in terms of the decomposition of Lemma 6.3.1, and in each case to consider just one summand.

(a) Let $\omega = f\, dx^{i_1} \wedge \cdots \wedge dx^{i_k}$ so that $I(\omega) = 0$. Then $d(I(\omega)) = 0$ also so we only need to consider $I(d\omega)$. However $d\omega = \omega_1 + \frac{\partial f}{\partial t} dt \wedge dx^{i_1} \wedge \cdots \wedge dx^{i_k}$, where ω_1 does not involve dt, so

$$\begin{aligned} I(d\omega)(p) &= \left(\int_0^1 \frac{\partial f}{\partial t}(p,t)dt\right) dx^{i_1} \wedge \cdots \wedge dx^{i_k} \\ &= \{f(p,1) - f(p,0)\}\, dx^{i_1} \wedge \cdots \wedge dx^{i_k} \\ &= \{f \circ \iota_1(p) - f \circ \iota_0(p)\}\, dx^{i_1} \wedge \cdots \wedge dx^{i_k} \\ &= \iota_1^*(\omega)(p) - \iota_0^*(\omega)(p) \end{aligned}$$

as now required, since $\iota_t^*(dx^{i_1} \wedge \cdots \wedge dx^{i_k}) = dx^{i_1} \wedge \cdots \wedge dx^{i_k}$.

(b) Let ω be a summand of $dt \wedge \omega_2$, that is, $\omega = f\, dt \wedge dx^{i_1} \wedge \cdots \wedge dx^{i_{k-1}}$. Then $\iota_1^*(\omega) = 0 = \iota_0^*(\omega)$, since $\iota_t^*(dt) = 0$ as we saw above. However

$$d\omega = \sum_{j=1}^m \frac{\partial f}{\partial x_j} dx^j \wedge dt \wedge dx^{i_1} \wedge \cdots \wedge dx^{i_{k-1}},$$

so that

$$I(d\omega) = -\sum_{j=1}^m \left(\int_0^1 \frac{\partial f}{\partial x_j} dt\right) dx^j \wedge dx^{i_1} \wedge \cdots \wedge dx^{i_{k-1}}.$$

On the other hand

$$\begin{aligned} d(I(\omega)) &= d\left\{\left(\int_0^1 f\, dt\right) dx^{i_1} \wedge \cdots \wedge dx^{i_{k-1}}\right\} \\ &= \sum_{j=1}^m \left(\int_0^1 \frac{\partial f}{\partial x_j} dt\right) dx^j \wedge dx^{i_1} \wedge \cdots \wedge dx^{i_{k-1}} = -I(d\omega), \end{aligned}$$

as required for this case, where we used the fact that

$$\frac{\partial}{\partial x_j}\left(\int_0^1 f\, dt\right) = \int_0^1 \frac{\partial f}{\partial x_j} dt$$

for a smooth function f. ■

We now see how this basic chain homotopy is transformed into a chain homotopy between the pull-backs of homotopic differentiable maps.

Theorem 6.3.3. *If f_0 and f_1 are homotopic (smooth) maps from N to M, then f_0^* and f_1^* are cochain homotopic pull-back cochain maps.*

Proof. Let $h : N \times \mathbb{R} \to M$ be a homotopy such that $h \circ \iota_t = f_t$ for $t = 0, 1$, and $I : \Omega^k(N \times \mathbb{R}) \to \Omega^{k-1}(N)$ be as in Lemma 6.3.2. Then we have the Diagram 6.3 from which we see that

$$\begin{aligned} f_1^*(\omega) - f_0^*(\omega) &= \iota_1^* \circ h^*(\omega) - \iota_0^* \circ h^*(\omega) \\ &= dI(h^*(\omega)) + I(d(h^*(\omega))) = d(I \circ h^*)(\omega) + (I \circ h^*)(d\omega). \end{aligned}$$

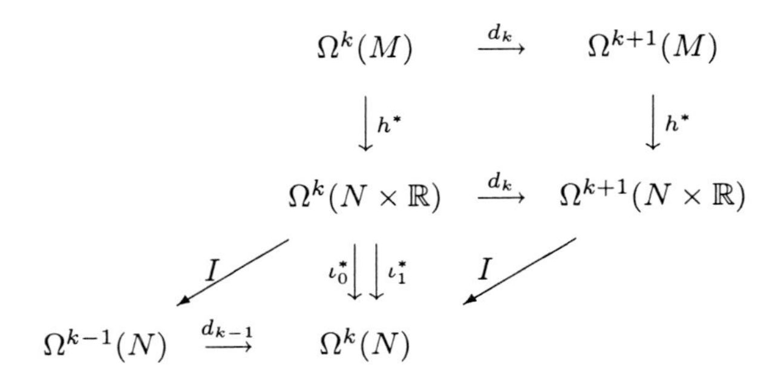

Diagram 6.3.

So $I \circ h^* : \Omega^*(M) \to \Omega^*(N)$, taking $\Omega^k(M)$ to $\Omega^{k-1}(N)$ is the required cochain homotopy between f_0^* and f_1^*. ■

Remark 6.3.4. It follows from our previous results that, if f_0 and f_1 are homotopic, then the linear maps of cohomology $(f_0^*)_*$ and $(f_1^*)_*$ induced by the pull-backs are identical. An important consequence is the following.

Corollary 6.3.5. *If M and N are homotopy equivalent manifolds, then $H^*(M) \cong H^*(N)$.*

Proof. Let $f : M \to N$ and $g : N \to M$ be inverse homotopy equivalences. Then $g \circ f \simeq \mathrm{id}_M$ implies that $(f^*)_* \circ (g^*)_* = (\mathrm{id}_M)_* = \mathrm{id}$, the identity of $H^*(M)$. Similarly $(g^*)_* \circ (f^*)_* = \mathrm{id}$, so that $(g^*)_*$ and $(f^*)_*$ are inverse isomorphisms between the cohomology spaces of M and those of N. ■

Examples 6.3.6. 1) Since $M \times \mathbb{R}^n \simeq M$ we have $H^k(M \times \mathbb{R}^n) \cong H^k(M)$ for all k.

2) A point is a zero-dimensional connected manifold. So $H^k(\{*\}) = 0$ for $k > 0$ and $H^0(\{*\}) = \mathbb{R}$. Hence the same is true of any contractible manifold. In particular, if M is compact and orientable with $\partial M = \emptyset$, then M is not contractible by Theorem 6.1.3.

3) We can think of our theorem, or more precisely its corollary, as a generalised Poincaré Lemma. However the classical version is the case of $\mathbb{R}^n$ in the previous example. This then says that $H^k(\mathbb{R}^n) = 0$ for $k > 0$ which, at the level of forms for which it is usually stated, says that any closed k-form is exact. For the record we give the proof directly from Lemma 6.3.2: let $h : \mathbb{R}^n \times \mathbb{R} \to \mathbb{R}^n$ be a contraction so that $h \circ \iota_0 =$ constant and $h \circ \iota_1 = \mathrm{id}_{\mathbb{R}^n}$. Then, for any closed k-form ω on $\mathbb{R}^n$,

$$(\iota_1^* - \iota_0^*)(h^*(\omega)) = dI(h^*(\omega)) - I(d(h^*(\omega))) = dI(h^*(\omega))$$

since $I(d(h^*(\omega))) = I(h^*(d\omega)) = 0$. But

$$(\iota_1^* - \iota_0^*)(h^*(\omega)) = (h \circ \iota_1)^*(\omega) - (h \circ \iota_0)^*(\omega) = \omega$$

since $(h \circ \iota_1)^* = (\mathrm{id}_{\mathbb{R}^n})^* = \mathrm{id}$, and $(h \circ \iota_0)^* = \mathrm{constant}^*(\omega) = 0$ since, for all v, $\mathrm{constant}_*(v) = 0$. Thus $\omega = dI(h^*(\omega))$ as claimed.

6.4. The Mayer-Vietoris Sequence.

The Poincaré Lemma, or more precisely the generalisation we have proved, will be one of our basic tools for calculating the de Rham groups of specific manifolds. The other will be the Mayer-Vietoris sequence which we now describe. We obtain it first at the level of forms and then for the cohomology groups. In each case the sequence is exact:

Definition 6.4.1. A sequence of vector spaces and linear maps is said to be *exact* if the image of each map is precisely the kernel of its successor.

In other words an exact sequence is a cochain complex all of whose cohomology groups vanish. The power of exact sequences in computations is exemplified by the following result for exact sequences of vector spaces, to which we shall have frequent recourse. It is an elementary consequence of exactness and the Rank-Nullity Theorem, which states that the rank plus the nullity of a linear mapping is equal to the dimension of the space on which it is defined.

Proposition 6.4.2. *In an exact sequence of vector spaces V_i and linear maps, $\theta_i : V_i \to V_{i+1}$, the alternating sum of the dimensions is zero: the exactness of the sequence*

$$0 \longrightarrow V_1 \longrightarrow V_2 \longrightarrow \cdots \longrightarrow V_k \longrightarrow 0$$

implies that $\sum_{i=1}^{k}(-1)^i \dim(V_i) = 0$.

Proof. This is an exercise in induction. For $k > 3$ we can break the long exact sequence into two shorter exact sequences:

$$0 \longrightarrow V_1 \longrightarrow V_2 \longrightarrow \cdots \longrightarrow V_{k-2} \longrightarrow \theta_{k-2}(V_{k-2}) \longrightarrow 0,$$

where we assume the result to hold by induction, and

$$0 \longrightarrow \ker(\theta_{k-1}) \longrightarrow V_{k-1} \longrightarrow V_k \longrightarrow 0,$$

where the result holds by the Rank-Nullity Theorem. These two cases combine to give the required result since $\theta_{k-2}(V_{k-2}) = \ker(\theta_{k-1})$ by the exactness of the original sequence. The induction may be started by the rank-nullity theorem at $k = 3$. ■

At the level of forms we shall have an exact sequence of cochain maps of cochain complexes of forms. By this we mean that the sequence of cochain maps restricts, for each degree k, to an exact sequence of linear maps between the relevant spaces of forms of that degree.

For an open subset U of M^m with inclusion $\iota : U \hookrightarrow M$ and a k-form ω on M, the pull-back $\iota^*(\omega)$ on U is called the *restriction of ω to U* since the coefficient functions are obtained by restricting those of ω to U.

Suppose that M is the union of two open subsets U and V and write $U \sqcup V$ for the disjoint union of two diffeomorphic copies of U and V. Then we have natural inclusions

$$\iota_U : U \cap V \longrightarrow U \sqcup V \quad \text{and} \quad \iota_V : U \cap V \longrightarrow U \sqcup V,$$

where points of $U \cap V$ are identified with their copies in U and V respectively, and also

$$\iota_M : U \sqcup V \longrightarrow M,$$

which maps the points of U and V to their images in M. We may identify $\Omega^*(U \sqcup V)$ with the graded direct sum $\Omega^*(U) \oplus \Omega^*(V)$.

Definition 6.4.3. With the data as above, the *Mayer-Vietoris sequence for forms on* (M, U, V) is

$$0 \longrightarrow \Omega^*(M) \xrightarrow{\iota_M^*} \Omega^*(U \sqcup V) \cong \Omega^*(U) \oplus \Omega^*(V) \xrightarrow{\iota_V^* - \iota_U^*} \Omega^*(U \cap V) \longrightarrow 0.$$

Note that, when we identify $\Omega^k(U \sqcup V)$ with $\Omega^k(U) \oplus \Omega^k(V)$ and write the general element of the latter as (ω_U, ω_V), the map $\iota_V^* - \iota_U^*$ maps this to $\iota_V^*(\omega_V) - \iota_U^*(\omega_U) = \omega_V\big|_{U\cap V} - \omega_U\big|_{U\cap V}$.

Theorem 6.4.4. *The Mayer-Vietoris sequence of forms that is associated with the triple* (M, U, V) *is graded exact. That is, it is the direct sum of the exact sequences*

$$0 \longrightarrow \Omega^k(M) \xrightarrow{\iota_M^*} \Omega^k(U) \oplus \Omega^k(V) \xrightarrow{\iota_V^* - \iota_U^*} \Omega^k(U \cap V) \longrightarrow 0.$$

Proof. Thinking through what it means for forms and their restrictions, exactness is clear except at $\Omega^k(U \cap V)$. Here, since everything is in the kernel of the zero map, exactness means that $\iota_V^* - \iota_U^*$ surjects.

To show that we use a partition of unity on M with just two functions ρ_U and ρ_V supported on U and V respectively. We may achieve that by modifying our previous construction by asking that each coordinate neighbourhood lie either in U or in V, allocating it arbitrarily if it lies in both. Then ρ_U is the sum of all the functions of the partition whose supports lie in coordinate neighbourhoods allocated to U, and ρ_V is defined similarly.

Then for $\omega \in \Omega^k(U \cap V)$ the form $-\rho_V\omega$ is zero on the open set $U \cap V \cap (\mathrm{supp}(\rho_V))^c = (U \setminus \mathrm{supp}(\rho_V)) \cap V$ so, defining it to be zero on $U \setminus \mathrm{supp}(\rho_V)$, it extends to a well-defined smooth form ω_U on U. Similarly $+\rho_U\omega$ extends to a form ω_V on V. Then, on restriction to $U \cap V$, $\omega_V - \omega_U$ becomes $\rho_U\omega - (-\rho_V\omega) = (\rho_U + \rho_V)\omega = \omega$. In other words, $(\iota_V^* - \iota_U^*)(\omega_U, \omega_V) = \iota_V^*(\omega_V) - \iota_U^*(\omega_U) = \omega$. ■

Theorem 6.4.5. *For* U *and* V *open in* $M^m = U \cup V$, *there is a long exact sequence, the Mayer-Vietoris sequence, for cohomology:*

$$\cdots \longrightarrow H^k(M) \xrightarrow{\imath} H^k(U) \oplus H^k(V) \xrightarrow{\jmath} H^k(U \cap V) \xrightarrow{\delta} H^{k+1}(M) \longrightarrow \cdots,$$

where the homomorphisms $\imath$, $\jmath$ *and* δ *will be defined below.*

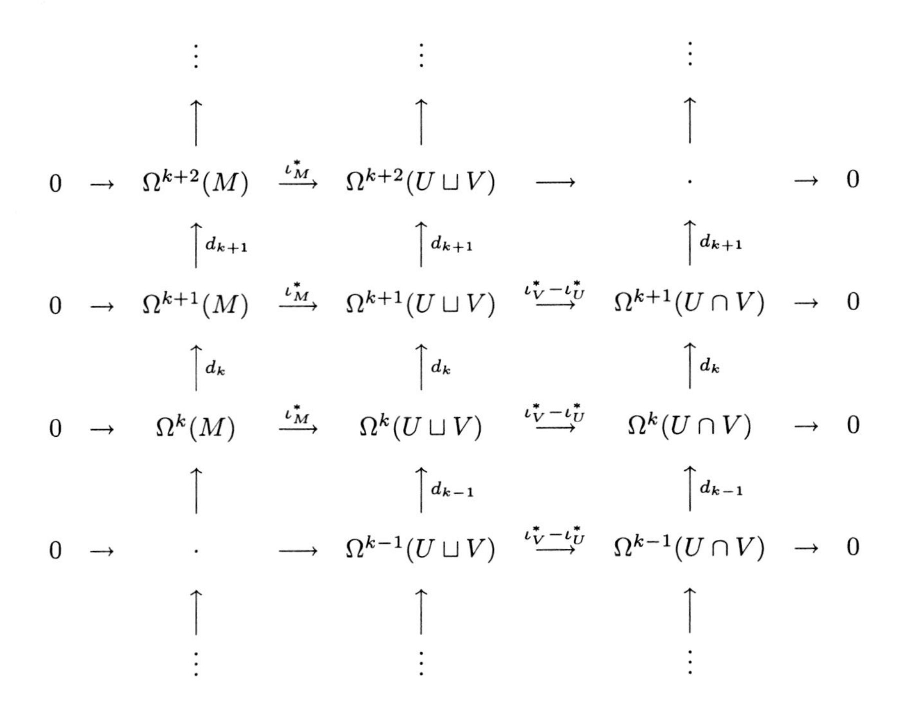

Diagram 6.4.

Proof. The Mayer-Vietoris sequence for forms gives Diagram 6.4 in which the rows are short exact sequences, the squares commute, that is for example $\iota_M^* \circ d = d \circ \iota_M^*$, and the columns are cochain complexes.

We note first that a closed form in $\Omega^*(U) \oplus \Omega^*(V)$ is a sum of one in U and one in V and similarly for exact forms. So the cohomology spaces arising from the cochain complex $\Omega^*(U) \oplus \Omega^*(V)$ form the (graded) direct sum $H^*(U) \oplus H^*(V)$. Hence we may define $\imath$ in the cohomology sequence to be the linear map $(\iota_M^*)_*$ induced by the pull-back ι_M^*. Similarly we define $\jmath$ to be $(\iota_V^* - \iota_U^*)_*$.

The definition of the *coboundary* δ in cohomology is more subtle and requires a 'chase' around Diagram 6.4. To aid perception we give the parallel Diagram 6.5 showing the location in the former diagram of all the elements to which we refer in the proof.

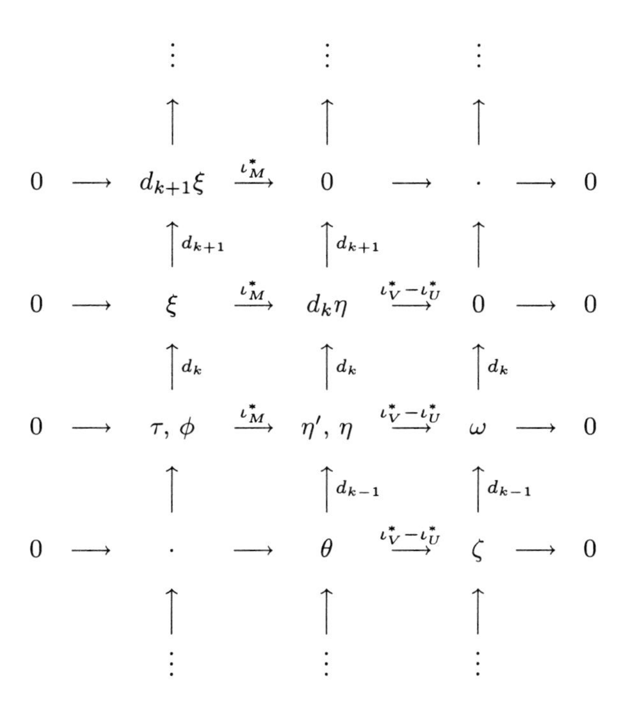

Diagram 6.5.

We represent a class $[\omega]$ in $H^k(U \cap V)$ by a closed k-form $\omega \in \Omega^k(U \cap V)$. Then, since $\iota_V^* - \iota_U^*$ surjects, there is a k-form η in $\Omega^k(U) \oplus \Omega^k(V)$ such that $(\iota_V^* - \iota_U^*)(\eta) = \omega$. Then $(\iota_V^* - \iota_U^*)(d_k\eta) = d_k(\iota_V^* - \iota_U^*)(\eta) = d_k\omega = 0$. So, by exactness of the horizontal short exact sequence at $\Omega^{k+1}(U) \oplus \Omega^{k+1}(V)$, there is a $(k+1)$-form ξ in $\Omega^{k+1}(M)$ such that $d_k\eta = \iota_M^*(\xi)$. Then

$$\iota_M^*(d_{k+1}\xi) = d_{k+1}\iota_M^*(\xi) = d_{k+1}d_k\eta = 0;$$

but ι_M^* is injective so $d_{k+1}\xi = 0$. Thus, since ξ is closed, we may define $\delta[\omega]$ to be the class $[\xi]$ in $H^{k+1}(M)$.

To see that $[\xi]$ is well-defined, we note first that ξ itself is already determined by η since ι_M^* is injective. If we now represent $[\omega]$ by $\omega' = \omega + d_{k-1}\zeta$ and choose an arbitrary $\eta' \in \Omega^k(U) \oplus \Omega^k(V)$ such that $(\iota_V^* - \iota_U^*)(\eta') = \omega'$, we may also choose $\theta \in \Omega^{k-1}(U) \oplus \Omega^{k-1}(V)$ such that $(\iota_V^* - \iota_U^*)(\theta) = \zeta$. Then $(\iota_V^* - \iota_U^*)(\eta' - \eta - d_{k-1}\theta) = d_{k-1}\zeta - (\iota_V^* - \iota_U^*)(d_{k-1}\theta) = 0$, since

$(\iota_V^* - \iota_U^*)(d_{k-1}\theta) = d_{k-1}(\iota_V^* - \iota_U^*)(\theta) = d_{k-1}\zeta$. So there is a φ in $\Omega^k(M)$ such that $\iota_M^*(\varphi) = \eta' - \eta - d_{k-1}\theta$. Then

$$\iota_M^*(\xi + d_k\varphi) = d_k\eta + d_k(\eta' - \eta - d_{k-1}\theta) = d_k\eta'.$$

Thus the alternative choices ω' and then η' lead to the same class $[\xi + d_k\varphi] = [\xi]$ in $H^{k+1}(M)$ as before.

We still have to prove that the long cohomology sequence is exact. To check this at $H^k(U \cap V)$, we compute $\delta[\omega]$ for the closed form ω in $\Omega^k(U \cap V)$ as above. Then $\delta[\omega] = 0$ means $[\xi] = 0$, that is, ξ is exact. Let $\xi = d_k\tau$ with τ in $\Omega^k(M)$. Then $d_k(\eta - \iota_M^*(\tau)) = d_k\eta - \iota_M^*(\xi) = d_k = 0$ and $(\iota_V^* - \iota_U^*)(\eta - \iota_M^*(\tau)) = \omega$, since $(\iota_V^* - \iota_U^*) \circ \iota_M^* = 0$. So $[\omega] = [(\iota_V^* - \iota_U^*)(\eta - \iota_M^*(\tau))] = \jmath[\eta - \iota_M^*(\tau)]$ as required.

The proofs of exactness at the other two stages of the cohomology sequence are similar exercises in 'diagram chasing' and are left to the reader. ■

We now have the tools ready to start calculating the cohomology spaces of various simple manifolds. We illustrate the technique by carrying out the calculation for spheres and tori in the next two sections.

6.5. The de Rham Groups of Spheres.

In the unit sphere $\mathbf{S}^m$ we let N and S denote the 'poles' $(\pm 1, 0, \cdots, 0)$ and $U = \mathbf{S}^m \setminus \{N\}$, $V = \mathbf{S}^m \setminus \{S\}$. Then $U \cup V = \mathbf{S}^m$ and $U \cap V \cong \mathbf{S}^{m-1} \times \mathbb{R} \simeq \mathbf{S}^{m-1}$. Also $U \cong V \cong \mathbb{R}^m$. The exact Mayer-Vietoris sequence determined by $(\mathbf{S}^m, U, V)$ starts with

$$\begin{aligned} 0 \longrightarrow H^0(\mathbf{S}^m) \longrightarrow H^0(U) \oplus H^0(V) \longrightarrow H^0(\mathbf{S}^{m-1}) \\ \longrightarrow H^1(\mathbf{S}^m) \longrightarrow H^1(U) \oplus H^1(V) \end{aligned}$$

which is equivalent to

$$0 \longrightarrow \mathbb{R} \longrightarrow \mathbb{R} \oplus \mathbb{R} \longrightarrow H^0(\mathbf{S}^{m-1}) \longrightarrow H^1(\mathbf{S}^m) \longrightarrow 0,$$

using our knowledge of the 0th de Rham space of a connected manifold and the fact that homotopy equivalent manifolds have isomorphic de Rham spaces in all degrees. Then, when $m = 1$, $\mathbf{S}^{m-1} = \mathbf{S}^0$ has two connected components so that $H^0(\mathbf{S}^0) \cong \mathbb{R} \oplus \mathbb{R}$ and, to make the alternating sum of dimensions zero, we must have $H^1(\mathbf{S}^1) \cong \mathbb{R}$. For $m > 1$, $\mathbf{S}^{m-1}$ is connected so $H^0(\mathbf{S}^{m-1}) \cong \mathbb{R}$ and we have $H^1(\mathbf{S}^m) = 0$. Continuing up the Mayer-Vietoris sequence we have, for $k \geqslant 2$, the sequence

$$H^{k-1}(U) \oplus H^{k-1}(V) \longrightarrow H^{k-1}(\mathbf{S}^{m-1}) \longrightarrow H^k(\mathbf{S}^m) \longrightarrow H^k(U) \oplus H^k(V),$$

in which the first and last entries are zero. Thus, for all $m \geqslant 1$ and $k \geqslant 2$,

$$H^{k-1}(\mathbf{S}^{m-1}) \cong H^k(\mathbf{S}^m),$$

which for $k > m$ gives $H^k(\mathbf{S}^m) \cong H^{k-m}(\mathbf{S}^0) = 0$, confirming what we know since there are no forms in degree k greater than the dimension of the manifold. When $k = m \geqslant 2$ we get $H^m(\mathbf{S}^m) \cong H^1(\mathbf{S}^1) \cong \mathbb{R}$ and, for $1 < k < m$, $H^k(\mathbf{S}^m) \cong H^1(\mathbf{S}^{m-k+1}) = 0$. Thus we have shown:

Theorem 6.5.1. *The de Rham cohomology groups of spheres $\mathbf{S}^m$ are all zero except for $H^0(\mathbf{S}^m) \cong \mathbb{R}$ and $H^m(\mathbf{S}^m) \cong \mathbb{R}$.*

6.6. The de Rham Groups of Tori.

By a torus we mean a product $T^n = \mathbf{S}^1 \times \cdots \times \mathbf{S}^1$ of n copies of the unit circle. The calculation of its cohomology groups is slightly less trivial than that for spheres. It proceeds by induction using the following theorem.

Theorem 6.6.1. *For $k \geqslant 0$, $H^{k+1}(M \times \mathbf{S}^1) \cong H^k(M) \oplus H^{k+1}(M)$.*

Remark 6.6.2. Of course we already know $H^0(M \times \mathbf{S}^1)$, since $M \times \mathbf{S}^1$ will have as many components as does M itself. Thus, since $H^{-1}(M)$ should clearly be taken to be zero as there are no (-1)-forms on M, the restriction to $k \geqslant 0$ is not strictly necessary. Similarly the following proof also re-calculates $H^{m+1}(M \times \mathbf{S}^1)$, which we could have found immediately if M and so also $M \times \mathbf{S}^1$, is orientable.

Proof. Let $U_N = M \times (\mathbf{S}^1 \setminus \{-i\})$ and $U_S = M \times (\mathbf{S}^1 \setminus \{i\})$. Both of U_N and U_S are open in $M \times \mathbf{S}^1$, with $U_N \cup U_S = M \times \mathbf{S}^1$, and $U_N \cap U_S$ is the union of two components $U_+ = M \times \{(x, y) \in \mathbf{S}^1 \mid x > 0\}$ and $U_- = M \times \{(x, y) \in \mathbf{S}^1 \mid x < 0\}$. Let $M_N = M \times \{i\}$, $M_S = M \times \{-i\}$, $M_+ = M \times \{+1\}$ and $M_- = M \times \{-1\}$. These sets are indicated symbolically in Figure 6.6.

Then the inclusions

$$M_N \xrightarrow{\iota_N} U_N, \quad M_S \xrightarrow{\iota_S} U_S, \quad M_+ \xrightarrow{\iota_+} U_+ \quad \text{and} \quad M_- \xrightarrow{\iota_-} U_-$$

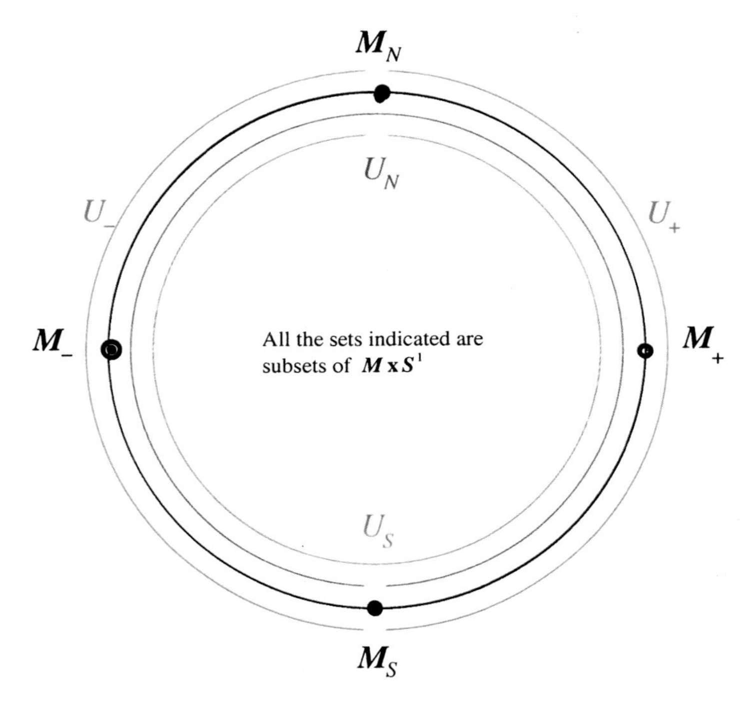

Figure 6.6.

are all homotopy equivalences and, since all the copies of M are diffeomorphic, they induce isomorphisms, for each k, of $H^k(M)$ with $H^k(U_N)$, $H^k(U_S)$, $H^k(U_+)$ and $H^k(U_-)$, respectively. See Figure 6.7.

We label the images under these isomorphisms of a general element x in $H^k(M)$ by x_N, x_S, x_+ and x_- respectively. Then, the inclusion $\iota_{+N} : U_+ \to U_N$ is homotopic to the composite of the inverse homotopy equivalence to the inclusion of M_+ in U_+, followed by the natural identification of M_+ with M and then with M_N and then followed by the inclusion of M_N in U_N:

$$\iota_{+N} : U_+ \overset{\iota_+}{\longleftarrow} M_+ \cong M \cong M_N \overset{\iota_N}{\longrightarrow} U_N.$$

It follows that $(\iota_{+N}^*)_*$ maps each element x_N of $H^*(U_N)$ to the corresponding copy x_+ of $H^*(U_+)$, with similar results for ι_{+S}, ι_{+N} and ι_{-S}.

Now the inclusion ι_N of $U_N \cap U_S = U_+ \sqcup U_-$ in $U_S \sqcup U_N$ via U_N restricts to ι_{+N} and ι_{-N}, respectively, on the two components. See Figure 6.8.

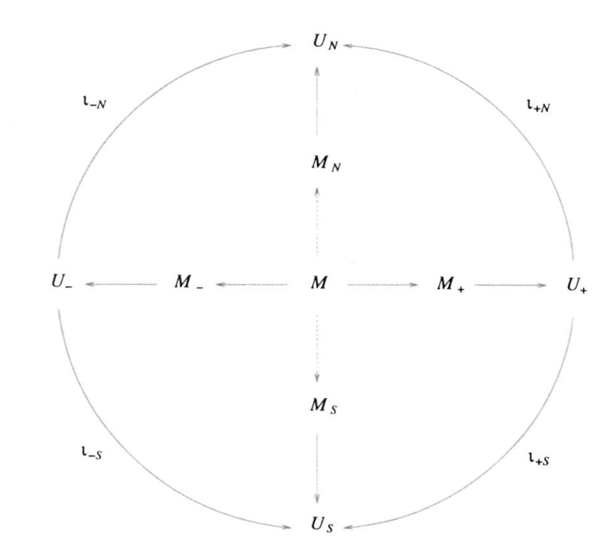

diffeomorphisms

inclusions which are homotopy equivalences

Figure 6.7.

Consequently

$$(\iota_N{}^*)_*(x_N) = \big((\iota_{+N}^*)_*(x_N), (\iota_{-N}^*)_*(x_N)\big) = (x_+,\, x_-),$$

with a similar computation for ι_S, the inclusion of $U_N \cap U_S$ in $U_S \sqcup U_N$ via U_S. Thus, since the linear map

$$H^k(U_N) \oplus H^k(U_S) \xrightarrow{\jmath} H^k(U_+) \oplus H^k(U_-)$$

in the Mayer-Vietoris sequence is induced by $(\iota_S{}^*)_* - (\iota_N{}^*)_*$, it is given by

$$\jmath : (x_N, y_S) \mapsto (y_+ - x_+,\, y_- - x_-) = ((y - x)_+,\, (y - x)_-)\,.$$

Thus it has kernel

$$\{(x_N, x_S) \mid x \in H^k(M)\} \cong H^k(M)$$

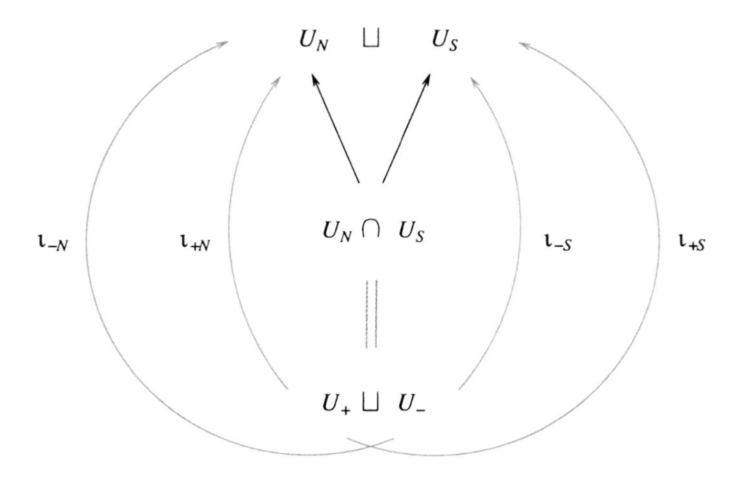

Figure 6.8.

and image

$$\{(x_+, x_-) \mid x \in H^k(M)\} \cong H^k(M).$$

Then the full Mayer-Vietoris sequence for cohomology gives the following short exact sequence:

$$0 \longrightarrow \ker(\delta) \longrightarrow H^k(U_N \cap U_S) \xrightarrow{\delta} H^{k+1}(M \times \mathbf{S}^1) \xrightarrow{\imath} \operatorname{im}(\imath) \longrightarrow 0,$$

in which $\ker(\delta) \cong \operatorname{im}(\jmath) \cong H^k(M)$, $\operatorname{im}(\imath) \cong \ker(\jmath) \cong H^{k+1}(M)$, from the long exact sequence and our calculation of $\jmath$, and

$$H^k(U_N \cap U_S) \cong H^k(U_+) \oplus H^k(U_-) \cong H^k(M) \oplus H^k(M).$$

Then the required result,

$$H^{k+1}(M \times \mathbf{S}^1) \cong H^k(M) \oplus H^{k+1}(M),$$

follows from the fact that the alternating sum of dimensions is zero in the exact sequence. ∎

Corollary 6.6.3. *The dimension of* $H^k(T^n)$ *is* $\binom{n}{k}$.

Proof. By induction, starting from the case $T^1 = \mathbf{S}^1$, which has already been proved, with the induction step being Theorem 6.6.1 with $M = T^n$:

$$\begin{aligned}\dim(H^{k+1}(T^{n+1})) &= \dim(H^k(T^n)) + \dim(H^{k+1}(T^n)) \\ &= \binom{n}{k} + \binom{n}{k+1} = \binom{n+1}{k+1}.\end{aligned}$$

■

Remark 6.6.4. The fact that $H^{k+1}(M \times \mathbf{S}^1)$ has an isomorphic copy of $H^{k+1}(M)$ as a direct summand is a general phenomenon for product manifolds. For, if

$$\iota : M \longrightarrow M \times N; \quad x \mapsto (x, *)$$

is the inclusion onto $M \times \{*\}$ where $*$ is some fixed point in N and

$$\pi_1 : M \times N \longrightarrow M; \quad (x, y) \mapsto x$$

is the projection onto the first factor, then

$$\iota^* \circ {\pi_1}^* = (\pi_1 \circ \iota)^* = (\mathrm{id}_M)^* = \mathrm{id}.$$

So, for each fixed k, $(\pi_1)_* : H^k(M) \to H^k(M \times N)$ is injective and, being a linear map between vector spaces, maps isomorphically onto a direct summand. Thus the special feature for the product $M \times \mathbf{S}^1$ is that the complementary subspace to $H^{k+1}(M)$ in $H^{k+1}(M \times \mathbf{S}^1)$ is isomorphic with $H^k(M)$. It is in fact the wedge product (in cohomology) of $({\pi_1}^*)_*(H^k(M))$ and $(\pi_2)_*(H^1(\mathbf{S}^1)) \cong \mathbb{R}$.

6.7. Homology and Submanifolds.

We conclude this chapter with a brief look at homology, the dual concept to cohomology. For many theories it is the homology that arises most naturally first and then the cohomology is defined to be its dual, rather than the other way around as it will be for us. As we have seen, cohomology behaves well for products, with the cohomology of factor manifolds being reflected in the cohomology of the product. For homology it is submanifolds that are more relevant. The duality arises as follows.

From the cochain complex Ω^*:

$$\cdots \longrightarrow \Omega^{k-1} \xrightarrow{d_{k-1}} \Omega^k \xrightarrow{d_k} \Omega^{k+1} \longrightarrow \cdots$$

we obtain the *chain* complex of dual spaces and dual maps:

$$\cdots \longleftarrow (\Omega^{k-1})^* \xleftarrow{\partial_k} (\Omega^k)^* \xleftarrow{\partial_{k+1}} (\Omega^{k+1})^* \longleftarrow \cdots$$

where $\partial_k = {d_{k-1}}^*$ and so $\partial_k \circ \partial_{k+1} = {d_{k-1}}^* \circ d_k^* = (d_k \circ d_{k-1})^* = 0$. Then the *kth homology group of* M is $\frac{\ker(\partial_k)}{\mathrm{im}(\partial_{k+1})}$ denoted by $H_k(M)$ or, to distinguish it from other homology groups, $H_k^{\mathrm{dR}}(M)$.

Proposition 6.7.1. $H_k(M) \cong (H^k(M))^*$.

Proof. $\mathrm{im}(\partial_{k+1}) = (\ker(d_k))^o$ and $\ker(\partial_k) = (\mathrm{im}(d_{k-1}))^o$ where, for $U \subseteq V$, the annihilator $U^o \subseteq V^*$ is $\{f \in V^* \mid f(U) = 0\}$. So

$$\frac{\ker(\partial_k)}{\mathrm{im}(\partial_{k+1})} = \frac{(\mathrm{im}(d_{k-1}))^o}{(\ker(d_k))^o} \cong \left(\frac{\ker(d_k)}{\mathrm{im}(d_{k-1})}\right)^* . \qquad \blacksquare$$

Proposition 6.7.2. *If* N^n *is a compact oriented manifold without boundary, then a differentiable map* $f : N \to M$ *determines a homology class* $[f] \in H_n(M)$.

Proof. The map

$$\phi : H^n(M) \longrightarrow \mathbb{R}; \quad [\omega] \mapsto \int_N f^*(\omega)$$

is well-defined since

$$\int_N f^*(\omega + d\eta) = \int_N f^*(\omega) + \int_N f^*(d\eta) = \phi([\omega]) + \int_{\partial N} \imath^* \circ f^*(\eta)$$

by Stokes' Theorem, which is $\phi([\omega])$ since $\partial N = \emptyset$. Also ϕ is linear since both f^* and integration over N are. Thus it does determine a class $[f] \in (H^n(M))^* = H_n(M)$. $\blacksquare$

Remark 6.7.3. In the special case when f is the embedding of a submanifold $N \hookrightarrow M$, we would denote its homology class by $[N] \in H_n(M)$. In general not every homology class is represented by a submanifold.

6.8. Exercises for Chapter 6.

All manifolds M^m and N^n may be assumed to be connected. Solutions to Exercises 6.1 and 6.2 should only involve forms, not cohomology. All cohomology is de Rham cohomology.

6.1. (*i*) Prove that a continuous map $f : M \to N$ may be approximated by a smooth one.
(*ii*) Let $f : M \to N$ and $g : N \to M$ be such that $f \circ g \simeq \mathrm{id}_N$. Prove that, if every closed k-form on M is exact, then the same is true on N.
(*iii*) Given the classical Poincaré Lemma, that every closed k-form on $\mathbb{R}^m$ with $k > 0$ is exact, deduce that, if M^m, $m > 0$, is a compact oriented manifold without boundary, then M is not contractible.

6.2. Use the Mayer-Vietoris sequence for forms (plus Question 6.1(*ii*) and the classical Poincaré Lemma) to show, for $1 < k < n$, that if every closed $(k-1)$-form on $\mathbf{S}^{n-1}$ is exact then every closed k-form on $\mathbf{S}^n$ is exact. Note which commutative squares you use; where you use exactness and which aspect, ker $\supseteq$ im or ker $\subseteq$ im, you need. Note also where you use the constraint on k.
Modify the proof to show that every closed 1-form on $\mathbf{S}^n$ is exact for $n > 1$.

6.3. Let U, V be open subsets of a manifold M such that $M = U \cup V$. Prove the exactness of the cohomology sequence

$$\longrightarrow H^k(M) \xrightarrow{\imath} H^k(U) \oplus H^k(V) \xrightarrow{\jmath} H^k(U \cap V) \xrightarrow{\delta} H^{k+1}(M) \longrightarrow$$

at $H^k(U) \oplus H^k(V)$ and at $H^k(M)$.

6.4. Let $f : \mathbb{R}^m \hookrightarrow M^m$ be an embedding, where M is compact, connected, orientable and without boundary and $m > 0$. Let $\mathring{M} = M \setminus f(\overline{B})$, where $\overline{B}$ is the closed unit ball in $\mathbb{R}^m$. Given that $H^m(\mathring{M}) = 0$, calculate the cohomology of $\mathring{M}$ in terms of that of M.

6.5. The connected sum $M\#N$ of two compact connected manifolds of dimension m may be defined using two embeddings $f_M : \mathbb{R}^m \hookrightarrow M$ and $f_N : \mathbb{R}^m \hookrightarrow N$ and on

$$W = (M \setminus \{f_M(0)\}) \sqcup (N \setminus \{f_N(0)\})$$

defining the equivalence relation $f_M(x) \sim f_N(x/\|x\|^2)$. [Points not in the image of f_M or f_N are equivalent only to themselves.] Then $M\#N$ is $W/\sim$ with the natural induced differential structure.

(i) Prove that $M\#N$ is orientable if and only if both M and N are.

(ii) If M and N are orientable and $m > 1$, calculate $H^*(M\#N)$ in terms of $H^*(M)$ and $H^*(N)$.

CHAPTER 7

DEGREES, INDICES AND RELATED TOPICS

This chapter is the culmination of those that comprise the underlying course, although in practice the lecturer only has time to select and briefly touch on a few of the topics we mention here. At some points, particularly in the final section, our exposition will be less explicit than in the earlier chapters. It is hoped that this will not be an impediment to understanding but rather be a spur to further study of the literature to which we refer.

The basic invariant that we shall use is the degree of a mapping between c.c.o. manifolds of the same dimension. In the first section we define this, show that it is an integer and show how it may be calculated from an analysis of the mapping near a regular value. In the next section we use the degree to define the linking number between submanifolds, or more generally maps of manifolds, of dimensions m and n in a Euclidean space of dimension $m+n+1$, as the degree of a related mapping from the product of the manifolds to the unit sphere in that space. This has particular interest since it is one of the very few examples of a geometric concept that, in the case that we may visualise explicitly, here circles in 3-space, is neither trivial nor atypical.

In section 3 we use the degree to define the index of a vector field, or more precisely the index at an isolated zero of that field. The local definition, for a vector field on an open subset of a Euclidean space is straightforward. However it requires a sequence of local lemmas to prove that it may be generalised to an unambiguous index for a vector field on a manifold and, incidentally, to see how it might be calculated. The most remarkable property of this index is that it is an invaraint, not just of the vector field, but also of the underlying manifold. In section 4 we give a first proof of this by showing that the index of any vector field is the degree of the Gauss map for an embedding of the manifold in a Euclidean space.

An alternate, eventually more revealing, explanation involves the Morse functions that we introduce in section 5. These are functions each of whose critical points is non-degenerate, and hence isolated, and for which we may also define an index using the Hessian matrix of second order partial derivatives at that point. Associated with a Morse function is a particular class of 'gradient-like' vector fields and the main result of this section is the theorem that such a, and hence by the result of the preceding section any, vector field has index equal to the alternating sum $\sum_{i=0}^{\dim}(-1)^k\lambda_k$, where λ_k is the number of critical points of index k. Conversely from a vector field with non-degenerate critical points one could construct a Morse function for which it is gradient-like.

The next section introduces the Euler number of a manifold, which is the similar alternating sum of the dimensions of its cohomology spaces and we establish the 'Poincaré–Hopf' theorem that this too is equal to the index of any vector field on the manifold. The proof proceeds by analysing the neighbourhood of each critical point of a Morse function and using the Mayer–Vietoris sequence to translate one alternating sum into the other. Analysing that neighbourhood more carefully we arrive at handle decompositions, which form the final topic in the chapter. In particular we show how a handle decomposition may be used to calculate the cohomology, so explaining the connection between the Euler number and the index of a gradient-like vector field associated with a Morse function. We conclude by identifying the de Rham cohomology groups with those of any other cohomology theory with real coefficents, both axiomatically and by explicit calculation involving the degrees of the attaching maps of the handles.

7.1. The Degree of a Mapping.

The fact that the cohomology in the top degree of a c.c.o. manifold-without-boundary is of dimension one leads immediately to the possibility of making the following definition.

Definition 7.1.1. If $f : M \to N$ is a smooth map between two compact connected oriented manifolds without boundary of the same dimension m, then $f^*([\omega_N]) = \deg(f)[\omega_M]$ defines the real number $\deg(f)$ called the *degree of f*. See Diagram 7.1.

If q is a regular value of f and $p \in f^{-1}(q)$, then by the Inverse Function Theorem, since M and N have the same dimension, there are neighbourhoods U of p and V of q such that f is a diffeomorphism of U onto V. We may also assume U and V sufficiently small to support charts ϕ and ψ respectively in the orientations of M and N. Then we define the *degree of f at p*, denoted by

$$\begin{array}{ccccccc}
[\omega_M] & \in & H^m(M) & \longleftarrow & H^m(N) & \ni & [\omega_N] \\
\downarrow & & \downarrow \cong & & \downarrow \cong & & \downarrow \\
1 & \in & \mathbb{R} & & \mathbb{R} & \ni & 1 \\
 & & \deg(f) & \longleftarrow & 1 & &
\end{array}$$

Diagram 7.1.

$I(f,p)$, to be $+1$ if $\det(J(\psi \circ f \circ \phi^{-1})) > 0$ and -1 if $\det(J(\psi \circ f \circ \phi^{-1})) < 0$. We also then say that f is *orientation preserving* or *reversing at* p respectively.

Note that a change of orientation of either M or N changes the sign of the degree of the map.

Theorem 7.1.2. *If q is a regular value of $f : M \to N$, where M and N are compact connected oriented manifolds without boundary, then*

$$\deg(f) = \sum_{p \in f^{-1}(q)} I(f,p).$$

Proof. Since q is a regular value, each point $p_i \in f^{-1}(q)$ has a coordinate neighbourhood U_i that is mapped diffeomorphically by f. Since M is compact there can only be finitely many such p_i and hence we can restrict the U_i such that each has the same image V, which we may also take to be a coordinate neighbourhood of q.

Using Proposition 5.5.4, we may represent the fundamental class of N by a form ω_N with $\operatorname{supp}(\omega_N) \subseteq V$. We also let ω_M represent the fundamental class of M, so that $f^*[\omega_N] = \deg(f)\,[\omega_M]$ implies $\int_M f^*(\omega_N) = \int_M \deg(f)\,\omega_M$ as the integral of any exact form over M is zero by Stokes' Theorem since $\partial M = \emptyset$. However

$$\operatorname{supp}(f^*(\omega_N)) \subseteq f^{-1}(V) = U_1 \cup \cdots \cup U_r,$$

the components being finite in number by the compactness of M again. So $f^*(\omega_N) = \sum_{i=1}^{r} \omega_i$ where $\omega_i \in \Omega^m(M)$ has $\operatorname{supp}(\omega_i) \subseteq U_i$ and $\omega_i\big|_{U_i}$ is given by

$(f|_{U_i})^*(\omega_N|_V)$. Thus

$$\begin{aligned}\deg(f) &= \deg(f)\int_M \omega_M = \int_M \deg(f)\,\omega_M = \int_M f^*(\omega_N) = \sum_{i=1}^r \int_M \omega_i \\ &= \sum_{i=1}^r \int_{U_i} (f|_{U_i})^*(\omega_N|_V) = \sum_{i=1}^r I(f,p_i)\int_V \omega_N|_V,\end{aligned}$$

since the effect of $(f|_{U_i})^*$ is just that of a change of coordinates which, if $I(f,p_i) = +1$, does not effect the value of the integral and, if $I(f,p_i) = -1$, simply changes the sign of the integral. However, since ω_N is supported on V,

$$\int_V \omega_N|_V = \int_N \omega_N = 1,$$

so that $\deg(f) = \sum_{i=1}^r I(f,p_i)$. ■

Remark 7.1.3. In particular the theorem shows that we may compute the degree of a map at any regular value. A classical theorem due to Sard says that regular values are dense in N. (See Chapter 9.)

Corollary 7.1.4. *The degree of f is an integer which only depends on the homotopy class of f.*

Proof. We have already seen that the induced map $(f^*)_*$ on cohomology depends only on the homotopy class. ■

Corollary 7.1.5. *If $g : N^m \to P^m$ is a map from N to a further manifold P of the same dimension, then* $\deg(g \circ f) = \deg(g)\deg(f)$.

Proof. For $\omega \in \Omega^m(P)$ we have

$$\begin{aligned}\deg(g\circ f)\int_P \omega &= \int_N (g\circ f)^*(\omega) = \int_N f^*(g^*(\omega)) \\ &= \deg(f)\int_M g^*(\omega) = \deg(f)\deg(g)\int_P \omega.\end{aligned}$$

■

7.2. Linking Numbers.

As a first application of the degree of a mapping, we use it to define the linking number between maps into a Euclidean space of two manifolds whose dimensions sume to one less than the dimension of the space.

Definition 7.2.1. Let M^m and N^n be two compact connected oriented manifolds without boundary and let $f : M \to \mathbb{R}^{m+n+1}$, $g : N \to \mathbb{R}^{m+n+1}$ be smooth maps such that $f(M) \cap g(N) = \emptyset$. Then the *linking number* $L(f,g)$ *of* f *and* g is the degree of the map

$$\Phi(f,g) : M \times N \longrightarrow \mathbf{S}^{m+n}; \quad (x,y) \mapsto \frac{g(y) - f(x)}{\|g(y) - f(x)\|}.$$

Special cases. 1) If M and N are submanifolds of $\mathbb{R}^{m+n+1}$ and $f = \iota_M$, $g = \iota_N$ the inclusions, then $L(\iota_M, \iota_N)$ is called the linking number $lk(M,N)$ of M and N.

2) If M and N are disjoint simple closed oriented curves in $\mathbf{S}^3$, or $\mathbb{R}^3$, $lk(M,N)$ is the classical linking number which is computed from a planar projection by assigning the number -1 to points of the projection where N passes over M with orientations as in Figure 7.2(i) and assigning the number $+1$ to points where N passes over M with orientations as in Figure 7.2(ii) and adding the results.

Figure 7.2.

To see that this is the case, imagine that the projection is onto the (x-y)-plane, with the unit sphere centered on the origin, and consider which points $p \in M$ and $q \in N$ are such that $\Phi(p,q)$ is close to the 'north pole' $(0,0,1)$. These must be the points near where N crosses over M. Taking p to be a point immediately under N, as q varies nearby in N the image $\Phi(p,q)$ traces out a line segment on $\mathbf{S}^2$ through the north pole. Now allowing p to vary in M, the corresponding lines trace out a (curvilinear) square on the surface of the sphere. This shows that the north pole is a regular value of Φ. Note however that, as we follow the positive orientation of M, to the right in (*i*), the resulting line segments move away from the original one in the opposite direction, to the left in this case. Thus, although the nearby orientations of M and N, taken in that order, project to the positive orientation of the plane, the map Φ has local degree -1 at (p, q). Similarly we see that it has degree $+1$ at points where N crosses as in (*ii*). The result follows from Theorem 7.1.2.

3) If $M = \mathbf{S}^1$ and N is a single point, then $f : \mathbf{S}^1 \to \mathbb{R}^2 \cong \mathbb{C}$ and $g(N)$ is a point $a \notin \operatorname{im}(f)$. Also $M \times N = \mathbf{S}^1$ and $\Phi(f,g)$ reduces to

$$\Delta(f) : \mathbf{S}^1 \longrightarrow \mathbf{S}^1; \quad z \mapsto \frac{f(z)-a}{\|f(z)-a\|},$$

where $z = e^{2\pi it}$, the degree of which is the *winding number* $W(f,a)$ of complex analysis.

Proposition 7.2.2. *Let $F : M \times \mathbb{R} \to \mathbb{R}^{m+n+1}$ and $G : N \times \mathbb{R} \to \mathbb{R}^{m+n+1}$ be homotopies of maps such that, for all t in $\mathbb{R}$, $F(M \times \{t\}) \cap G(N \times \{t\}) = \emptyset$. Then*

$$L(f_0, g_0) = L(f_1, g_1),$$

where $f_t(x) = F(x,t)$ and $g_t(y) = G(y,t)$.

Proof. We use F and G to define a homotopy between $\Phi(f_0, g_0)$ and $\Phi(f_1, g_1)$. ∎

Corollary 7.2.3. *If f and g are homotopic maps of $\mathbf{S}^1$ into $\mathbb{C} \setminus \{a\}$, then $W(f,a) = W(g,a)$.* ∎

Corollary 7.2.4. *If $f(M)$ and $g(N)$ lie on opposite sides of a hyperplane in $\mathbb{R}^{m+n+1}$, then $L(f,g) = 0$.*

Proof. We may find totally disjoint homotopies of the two maps to constant maps. ∎

7.3. The Index of a Vector Field.

Firstly, as usual, we deal with the local case. Let X be a vector field on an open subset U of $\mathbb{R}^m$. Recall that, since the tangent bundle to $\mathbb{R}^m$ is trivial, we may think of X as a mapping from U into $\mathbb{R}^m$, where each $X(x)$ for $x \in U$ is regarded as a tangent vector to U at x. As already mentioned in Chapter 2, a convenient way to visualise a vector field on a Euclidean space, at least in low dimensions, is to draw the vector $X(x_i)$ at a sufficiently typical set of points x_i. Alternatively one can draw a range of integral curves. Recall that an integral curve of X is a curve $\gamma : (a, b) \to U$, where $(a, b) \subset \mathbb{R}$, such that, for all $t \in (a, b)$, $\gamma'(t) = X(\gamma(t))$. Note however that, for points on each integral curve drawn, this latter picture only indicates the direction of $X(\gamma(t))$, not its magnitude. In Figure 7.3 we illustrate the integral curves of a familiar vector field, the magnetic field of a bar magnet, in the plane.

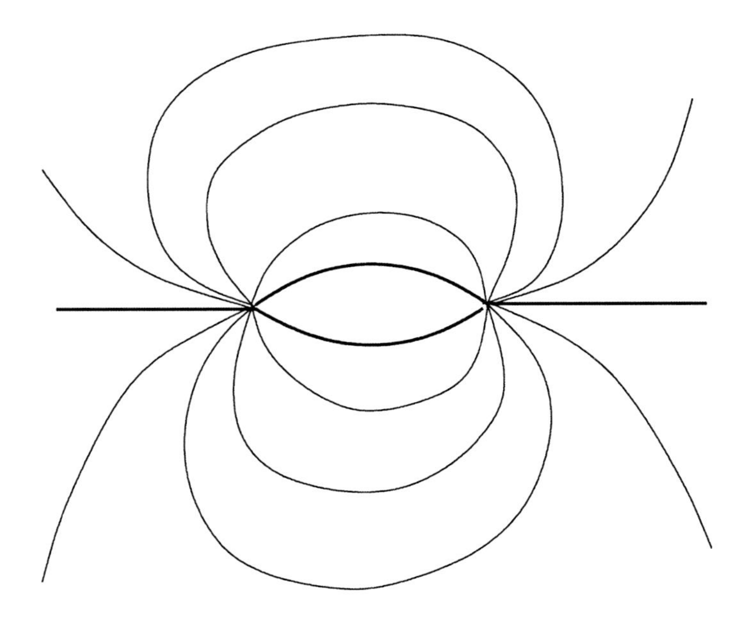

Figure 7.3.

A zero of the vector field is also referred to as a *singularity*. We can only deal with isolated singularities, but that will involve no true loss of generality. For an isolated singularity, p, of X choose ρ such that the ball $B(p, \rho)$ is contained in U and p is the only singularity of X in $B(p, \rho)$. Then we may define

$$f_\epsilon : \mathbf{S}^{m-1} \longrightarrow \mathbf{S}^{m-1}; \quad x \mapsto \frac{X(p+\epsilon x)}{\|X(p+\epsilon x)\|},$$

for $0 < \epsilon < \rho$, since $p + \epsilon x \in U$ for all $x \in \mathbf{S}^{m-1}$, and there is an obvious homotopy between any two such maps. Thus, by Corollary 7.1.4, $\deg(f_\epsilon)$ is also independent of ϵ for $0 < \epsilon < \rho$. It is called the *index of X at p* and we shall denote it by $I(X, p)$. For example, each of the singularities in Figure 7.3 has index 1, irrespective of which way the field, and hence its integral curves, is coherently oriented: both X and $-X$ have the same index at each critical point. That is a property common to all vector fields on even dimensional manifolds. In Figure 7.4 we illustrate the integral curves of a vector field in the plane with a single critical point of index 2. Note that, sufficiently far from the critical points, the two fields illustrated in Figure 7.3 and Figure 7.4 are very similar.

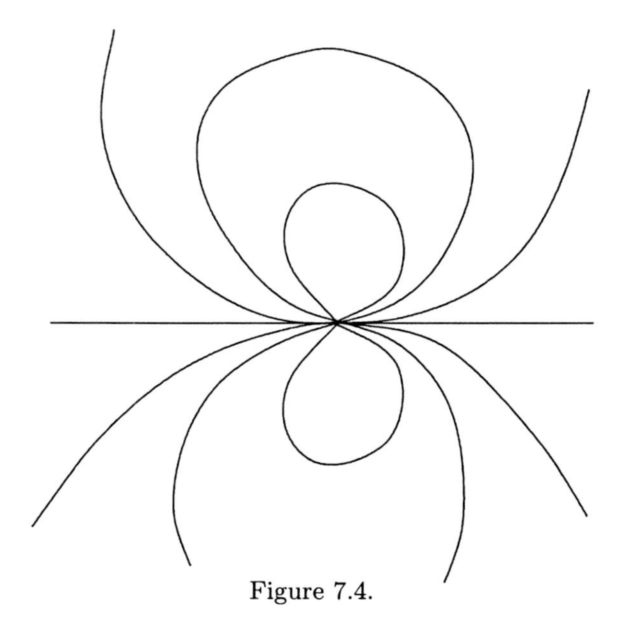

Figure 7.4.

In order to extend this definition of the index to vector fields on manifolds we need to examine a few of its properties in Euclidean spaces. Firstly we note that it only depends on the 'germ' of X:

Proposition 7.3.1. *Let $X : U \to \mathbb{R}^m$ and $Y : V \to \mathbb{R}^m$ be two vector fields whose only singularity is at the origin and such that, for some $\delta > 0$, $X|_B = Y|_B$, where $B \equiv B(0, \delta)$, the ball about the origin of radius δ. Then $I(X, 0) = I(Y, 0)$.*

Proof. This is immediate from the fact that ϵ may be arbitrarily small when defining the map f_ϵ. ■

If the ball $B \equiv B(0, 2\delta)$ is contained in U and ϕ is a diffeomorphism of B with $\mathbb{R}^m$ which is the identity on $B(0, \delta)$, then $X \circ \phi^{-1}$ is a vector field with the same germ, and hence the same index, at the origin as X. Thus, when discussing the indices of vector fields with their only singularities being at, say, the origin, we may assume they are globally defined. We shall use that simplification in stating and proving the next results.

Proposition 7.3.2. *Let $X : \mathbb{R}^m \to \mathbb{R}^m$ and $Y : \mathbb{R}^m \to \mathbb{R}^m$ be two vector fields whose only singularity is at the origin. Then $Y \circ X$ is also a vector field whose only singularity is at the origin and $I(Y \circ X, 0) = I(Y, 0)\, I(X, 0)$.*

Proof. Let $\iota : \mathbf{S}^{m-1} \to \mathbb{R}^m \setminus \{0\}$ be the inclusion of the unit sphere and $\rho : \mathbb{R}^m \setminus \{0\} \to \mathbf{S}^{m-1}; x \mapsto x/\|x\|$ be the projection onto the unit sphere. Then $\iota \circ \rho$ is homotopic to the identity of $\mathbb{R}^m \setminus \{0\}$ and so $\rho \circ Y \circ X \circ \iota$ is homotopic to $\rho \circ Y \circ \iota \circ \rho \circ X \circ \iota$. Then

$$\begin{aligned} I(Y \circ X,\, 0) &= \deg(\rho \circ Y \circ X \circ \iota) = \deg(\rho \circ Y \circ \iota \circ \rho \circ X \circ \iota) \\ &= \deg(\rho \circ Y \circ \iota) \deg(\rho \circ X \circ \iota) = I(Y,\, 0)\, I(X,\, 0), \end{aligned}$$

by Corollary 7.1.5. ■

Proposition 7.3.3. *Let $X : \mathbb{R}^m \to \mathbb{R}^m$ and $Y : \mathbb{R}^m \to \mathbb{R}^m$ be two vector fields whose only singularity is at the origin and let $H : \mathbb{R}^m \times I \to \mathbb{R}^m$ be a homotopy between X and Y, fixing the origin. Then $I(X, 0) = I(Y, 0)$.*

Proof. The composition $\rho \circ H \circ (\iota \times \mathrm{id}_{\mathbb{R}^m})$ is a homotopy between $\rho \circ Y \circ \iota$ and $\rho \circ X \circ \iota$. ■

Proposition 7.3.4. *Let $X : \mathbb{R}^m \to \mathbb{R}^m$ be a linear isomorphism. Then, regarded as a vector field, $I(X,\, 0) = \mathrm{sign}(\det(X))$.*

Proof. A linear isomorphism certainly has its only singularity at the origin. It may be written as the composition of a (positive) magnification, to achieve a determinant equal to ± 1, and then shears and reflections. A magnification clearly has index +1 and a shear is homotopic to the identity (fixing the origin) and so it also has index +1. A reflection restricts to a reflection on the unit sphere, which is a map of degree −1. Thus a reflection of $\mathbb{R}^m$, regarded as a vector field has index −1. Since shears have determinant +1 and reflections determinant −1, the result follows by Proposition 7.3.2. ■

Proposition 7.3.5. *Let $X : \mathbb{R}^m \to \mathbb{R}^m$ be a vector field, whose only singularity is at the origin and for which $DX(0)$ is non-singular. Then $I(X, 0) =$* $\mathrm{sign}(\det(DX(0)))$.

Proof. This is a consequence of Propositions 7.3.3 and 7.3.4 using the following homotopy between X and its derivative $DX(0)$ at the origin.

$$H : \mathbb{R}^m \times \mathbb{R} \longrightarrow \mathbb{R}^m; \quad (x, t) \mapsto \begin{cases} DX(0)\cdot x & \text{if } t = 0, \\ X(tx)/t & \text{if } t \neq 0. \end{cases}$$

To see that this function H is smooth at $t = 0$ we note that, since $X(0) = 0$,

$$X(x) = \int_0^1 \frac{d}{dt} X(tx)\, dt = \int_0^1 \left(\sum_{i=1}^m x_i \frac{\partial X}{\partial x_i}(tx) \right) dt = \sum_{i=1}^m x_i X_i(x),$$

where $X_i \in \mathcal{C}^\infty(\mathbb{R}^m, \mathbb{R}^m)$ is defined by $X_i(x) = \int_0^1 \left(\frac{\partial X}{\partial x_i}(tx) \right) dt$. Then

$$H(x, t) = \sum_{i=1}^m x_i X_i(tx)$$

in both cases and so H is also smooth. ■

A singularity such as that in Proposition 7.3.5 is called *non-degenerate.* The following shows that, at the cost of increasing the number of isolated singularities, we can assume all singularities are non-degenerate without changing the index. This is a first glimpse of a more general theorem that says that any two vector fields on a given manifold have the same index.

Proposition 7.3.6. *Let $X : \mathbb{R}^m \to \mathbb{R}^m$ be a vector field, whose only singularity is at the origin. Then there is a vector field $Y : \mathbb{R}^m \to \mathbb{R}^m$, agreeing with X outside the ball of radius 2ϵ centred on the origin, that has only isolated non-degenerate singularities.*

Proof. Let ϕ be a bump function on $\mathbb{R}^m$ supported on $B(0, 2\epsilon)$ and identically 1 on $B(0, \epsilon)$. We define $Y(x) = X(x) - \phi(x)v$ and choose $v \in B(0, \delta)$ where

$\delta = \inf\{ \| X(x) \| \mid \epsilon \leqslant \| x \| \leqslant 2\epsilon\}$. Then all the zeros of Y lie in $B(0, \epsilon)$ on which $Y(x) = X(x) - v$. Choosing v to be a regular value of X, as we may by Sard's Theorem, gives Y the required properties. ■

We turn now to the global definition: the index of a vector field on a manifold. Before doing so we should recall that we have simplified our notation above for vector fields on Euclidean spaces, taking advantage of the triviality of the tangent bundle: if $X : \mathbb{R}^m \to \mathbb{R}^m$ is given by $X(x) = (X_1(x), \cdots, X_m(x))$ then, strictly, X is the vector field that, at x, is equal to $\sum_{i=1}^{m} X_i(x)\frac{\partial}{\partial x_i}$, where $\frac{\partial}{\partial x_1}, \cdots, \frac{\partial}{\partial x_m}$ is the standard basis of tangent vectors at x. Thus, if x is an isolated singularity of X, then Proposition 7.3.5 says that $I(X, x) = \text{sign}\left(\det\left(\frac{\partial X_i}{\partial x_j}\right)\right)$.

Note that, given a diffeomorphism $\phi : U \to V$ between open subsets of $\mathbb{R}^m$ or, more generally, between open subsets of two manifolds of dimension m, and a vector field X on U, we may define an induced vector field $\phi_*(X)$ on V, sometimes called the 'push-forward' of X by

$$\phi_*(X)(q) = \phi_*(p)(X(p)),$$

where p is the unique point $\phi^{-1}(q)$. Thus this definition works more generally for any injective mapping ϕ, in which case the vector fields X and $\phi_*(X)$ are said to be ϕ-related. However we shall only require it for coordinate charts and coordinate transformations.

Definition 7.3.7. Let X be a vector field, with only finitely many singularities $p_1, \cdots, p_k$, on the oriented manifold M. Then the *index* $I(X)$ *of* X is defined to be the sum

$$\sum_{i=1}^{k} I((\phi_i)_*(X),\ \phi_i(p_i)),$$

where each ϕ_i is a chart of the orientation defined on a neighbourhood containing p_i. The summand $I((\phi_i)_*(X),\ \phi_i(p_i))$ would also be called the *index of* X *at* p_i and be simply denoted $I(X,\ p_i)$, suppressing explicit mention of the coordinate chart, provided that that does not lead to confusion.

To see that this index is well-defined, we need the following result to show that it does not matter which charts we choose at each of the singularities. Note that, since the singularities are finite in number, they are necessarily isolated. We may also consider just one singularity and assume that one of the charts we choose there is a standard chart $\psi : V \to \mathbb{R}^m$.

Lemma 7.3.8. *Let a vector field X be given on an open subset U of $\mathbb{R}^m$ with just one singularity p and let $\phi : U \to V$ be a diffeomorphism. Then*

$$I(X, p) = I(\phi_*(X), \phi(p)).$$

Proof. Clearly we may assume that $p = 0$ and $\phi(p) = 0$, so that both X and $\phi_*(X)$ have the point 0 as their only singularity.

As a map from V to $\mathbb{R}^m$, $\phi_*(X)$ is given by

$$v \mapsto D\phi(\phi^{-1}(v)) \cdot X(\phi^{-1}(v)).$$

So it is the composition of ϕ^{-1} followed by the map $Y : u \mapsto D\phi(u) \cdot X(u)$. Now $H : U \times \mathbb{R}; (u, t) \mapsto D\phi(tu) \cdot X(u)$ is a homotopy, fixing the origin, between Y and $D\phi(0) \circ X$. Thus the lemma follows from the preceding propositions. ■

7.4. The Gauss Map.

The index of a vector field on a manifold has several apparently very different interpretations. In this section we shall interpret it using the Gauss map. In particular this will show that the index is an invariant of the manifold: all vector fields with isolated singularites on a given manifold have the same index.

For an oriented hyper-surface M^m embedded in $\mathbb{R}^{m+1}$ the orientation of M determines, at each point p of M, a choice of unit normal $\nu(p)$ by requiring that coordinates from the orientation of M at p, followed by an $(m+1)$st coordinate axis directed along $\nu(p)$ gives a chart in the standard orientation of $\mathbb{R}^{m+1}$ at p. Then the *Gauss map* is

$$G : M \longrightarrow \mathbf{S}^m; \quad p \mapsto \nu(p),$$

where, as usual, we suppress explicit mention of the implied parallel translation of each unit vector in $\tau_p(M)$ to the tangent space at the origin and the identification of this tangent space with $\mathbb{R}^{m+1}$ itself.

If, now, M is the boundary of an $(m+1)$-submanifold-with-boundary N of $\mathbb{R}^{m+1}$, then N inherits an orientation from the standard orientation of $\mathbb{R}^{m+1}$ and, from that, M inherits an orientation as its boundary. The resulting normal field along M obtained above will everywhere point 'out' of N. Then the degree of the Gauss map of M is equal to the index of any vector field on N, provided that its singularities are isolated and none lies on M and that along

M it points out of N. Note that for this result neither N nor its boundary M need be connected, though for the proof it is sufficient to assume that N is connected.

In order to prove this result, and also to extend it to general submanifolds of $\mathbb{R}^n$ and hence, via Whitney's Theorem, to arbitrary abstract manifolds, we shall need a technical tool: one that has many other applications in the theory of manifolds.

Lemma 7.4.1. *If M^m is a compact submanifold of $\mathbb{R}^n$, then there is an open neighbourhood T of M in $\mathbb{R}^n$ that has the structure of a vector bundle over M, with fibre dimension $k = n - m$, and for which the inclusion of M in T is the zero cross-section.*

Proof. Consider the subset $N = \{(x, v) \mid x \in M,\, v \in \tau_x(M)^\perp\}$ of $\mathbb{R}^n \times \mathbb{R}^n$, where we regard $\tau_x(M)$ as a subspace of $\tau_x(\mathbb{R}^n)$ and $\tau_x(M)^\perp$ is its orthogonal complement. We show first that N is the total space of a vector bundle. At each point $p \in M$ we may take a chart (ϕ, U) at p on $\mathbb{R}^n$ adapted to M, with coordinates $x_1, \cdots, x_n$ on $\phi(U) \subset \mathbb{R}^n$. Then the tangent vector fields $\partial_i \equiv \frac{\partial}{\partial x_i}$ on $\phi(U)$ give a basis $\{(\phi^{-1})_*(q)(\partial_i) \mid i = 1, \cdots, m\}$ of $\tau_q(M)$ for each $q \in U$ so that $\{(\phi^{-1})_*(q)(\partial_i) \mid i = m+1, \cdots, n\}$ is a basis of the complement of $\tau_q(M)$ in $\tau_q(\mathbb{R}^n)$. Then the Gram-Schmidt orthonormalisation process, carried out simultaneously in all the tangent spaces $\tau_q(\mathbb{R}^n)$, for $q \in U$, will, since it then has smooth coefficient functions on U, produce new vector fields on U which are orthonormal at every point. Since the procedure preserves the flag of subspaces, that is, at each stage the new ith vector is in the space spanned by the first i original ones, the final k vector fields $e_1, \cdots, e_k$ will form an orthonormal basis of $\tau_q(M)^\perp$ at each $q \in U$. Then, if $\pi : N \to M$ is projection onto the first factor, the map

$$\Phi : \pi^{-1}(U \cap M) \longrightarrow \mathbb{R}^m \times \mathbb{R}^k; \quad (q,\, v) \mapsto (\phi(q),\, E(q)(v)),$$

where $E(q)(v) = (v \cdot e_1(q), \cdots, v \cdot e_k(q))$, gives a smooth local trivialisation of N which is linear on each fibre and so confirms that N is the total space of a vector bundle over M with bundle projection π.

We now let N_ϵ be the set $\{(q,\, v) \in N \mid \|v\| \leqslant \epsilon\}$ in N and define the map

$$f_\epsilon : N_\epsilon \longrightarrow \mathbb{R}^n; \quad (q,\, v) \mapsto q + v.$$

This clearly identifies the zero section $\{(q,\, 0) \in N \mid q \in M\}$ naturally with M and, in terms of the chart Φ above on $\pi^{-1}(U \cap M)$ and the identity chart on

$\mathbb{R}^n$, at such points its derivative is $D\phi^{-1}(\phi(q)) + E^{-1}(\phi(q))$. Since the former of these summands maps onto the tangent space at q and the latter onto the normal space it follows that this derivative is an isomorphism and hence that f_ϵ is a local diffeomorphism along the zero section. Since M is compact we may choose ϵ sufficiently small that f_ϵ is everywhere a local diffeomorphism. We may also choose ϵ so that f_ϵ is injective. For otherwise we could find sequences (p_n, u_n), (q_n, v_n) in $N_{1/n}$ such that

$$f_{1/n}(p_n,\, u_n) = p_n + u_n = q_n + v_n = f_{1/n}(q_n,\, v_n)$$

for all n. Taking subsequences if necessary, we may assume that $p_n \to p$ and $q_n \to q$ in M. Then, since $u_n \to 0$ and $v_n \to 0$, we must have $p = q$ and so a sequence of pairs (p_n, u_n), (q_n, v_n) of points in N_1 which converge to $(p,\, 0)$ and whose images coincide under f_1, contradicting the local injectivity of f_1 at $(p,\, 0)$. Then, since N_ϵ is compact and f_ϵ a continuous bijection, the latter is a homeomorphism onto its image. Thus f_ϵ maps the interior of N_ϵ, the set $\{(q,\, v) \in N \mid \|v\| < \epsilon\}$, onto an open submanifold T of $\mathbb{R}^n$.

To establish the properties of T specified in the statement of the lemma, we need only precede the diffeomorphism f_ϵ by a diffeomorphism of N onto $\mathrm{int}(N_\epsilon)$ preserving the zero-section. This is easily achieved by the map

$$(x,\, v) \mapsto (x,\, \lambda(v)v),$$

where $\lambda : [0,\, \infty) \to [0,\, \epsilon)$ is a diffeomorphism that restricts to the identity near the origin. ■

Remark 7.4.2. This lemma is by no means the strongest result of this nature that could be proved. Firstly M need not be compact: we then need to allow ϵ to be a function of $x \in M$. Secondly if M were embedded in an arbitrary manifold N we should still obtain such a neighbourhood of M in N. However the result we have proved is adequate for our immediate purposes. In particular, it has the two corollaries below that we shall use to interpret the Gauss map.

The vector bundle N is the *normal bundle* $\nu(M)$ of M in $\mathbb{R}^n$ that we met in Chapter 3. It is often denoted by $\nu\,(M \subset \mathbb{R}^n)$ since it is not an invariant of M alone, but depends also on its embedding. The image T of its embedding in $\mathbb{R}^n$ is called a *tubular neighbourhood of M in* $\mathbb{R}^n$ and is clearly non-unique. In many applications, including some below, the full vector bundle structure of the tubular neighbourhood is not required and it is also more convenient to work with a closed neighbourhood, such as $T_\epsilon(M) = f_\epsilon(N_\epsilon)$ in the above notation. In this case the tubular neighbourhood is diffeomorphic with the total space N_ϵ of the disc-bundle associated with the normal bundle. Its boundary

$$S_\epsilon(M) = f_\epsilon(\partial N_\epsilon), \qquad \text{where } \partial N_\epsilon = \{(x, v) \in N \mid \|v\| = \epsilon\},$$

is diffeomorphic with the total space of the associated sphere-bundle.

Corollary 7.4.3. *If M^m is a compact submanifold of $\mathbb{R}^{m+1}$ and X is a vector field in $\mathbb{R}^{m+1}$ defined along M that is nowhere tangent to M, then the inclusion of $M \equiv M \times \{0\}$ in $\mathbb{R}^{m+1}$ extends to an embedding of $M \times I$ in $\mathbb{R}^{m+1}$, where I is the open interval* $(-1, 1)$.

Proof. Since X is nowhere tangent to M it is also never-zero. Then, using $X(p)$ as the complement to $\tau_p(M)$ in the proof of the theorem, we obtain a unit normal field to M and so prove that the 1-dimensional bundle N is trivial. Hence, in this case, T is a product as stated. ■

Corollary 7.4.4. *If M^m is a compact submanifold of $\mathbb{R}^n$, the boundary $S_\epsilon(M)$ of a closed tubular neighbourhood of M is an $(n-1)$-dimensional compact submanifold of $\mathbb{R}^n$.*

Proof. Clearly ∂N_ϵ is a codimension one compact submanifold of N. Hence its image under the embedding into $\mathbb{R}^n$ is also a codimension one compact submanifold. ■

Theorem 7.4.5. *Let P be an oriented $(m+1)-$submanifold of $\mathbb{R}^{m+1}$ with boundary M and U be a domain in $\mathbb{R}^{m+1}$ containing P. Let X be a vector field on U with isolated singularities, all lying on P, and such that, for $x \in M$, $X(x)$ points out of P. Then, if G is the Gauss map of M determined by the orientation induced from that of P,*

$$I(X) = \deg(G).$$

Proof. To say that $X(x)$ points out of P means that, in terms of the standard coordinates we used for manifolds with boundary, for points x in M the first component of $X(x)$ is positive. In particular $X(x)$ is non-zero and does not lie in $\tau_x(M)$. Thus we may use the 'collar' of Corollary 7.4.3 and a bump function on $[-1, 1]$ to replace X by a homotopic field, which we shall still denote by X, that differs from X only near M, and in particular not on a neighbourhood of each of the singularities, and which on M is the unit outward normal.

Let $p_1, \cdots, p_k$ be the singularities of X and let $B_1, \cdots, B_k$ be a set of disjoint closed $(m+1)$-balls in $P \setminus M$ such that $p_i \in B_i$, for each i. On the compact manifold with boundary $Q = P \setminus \left(\bigcup_{i=1}^{k} \mathrm{int}(B_i) \right)$, since X is never zero there, we can define the function $F : Q \to \mathbf{S}^m;\ x \mapsto X(x)/\|X(x)\|$. Note that $F\big|_M$ is the Gauss map and that, if $f_i = F\big|_{\partial B_i}$, the restriction of F to the boundary sphere of the ith ball, then $\deg(f_i) = I(X, p_i)$ where, for the latter, ∂B_i is oriented as the boundary of B_i with the orientation induced from $\mathbb{R}^{m+1}$, However, if $\omega \in \Omega(\mathbf{S}^m)$ is such that $\int_{\mathbf{S}^m} \omega = 1$, then

$$\deg(F\big|_{\partial Q}) = \int_{\partial Q} F^*(\omega) = \int_Q d(F^*(\omega)) = \int_Q F^*(d\omega) = 0.$$

But $\deg(F\big|_{\partial Q}) = \deg(G) - \sum_i \deg(f_i)$, since the sphere ∂B_i has the opposite orientation as a boundary component of Q to that which it has as the boundary of B_i. The result follows. ■

Theorem 7.4.6. *Let M be a compact oriented m-manifold embedded in $\mathbb{R}^n$ and let $S_\epsilon(M)$ be the boundary of a closed tubular neighbourhood $T_\epsilon(M)$ of M. Let G be the Gauss map of $S_\epsilon(M)$, with respect to the normal pointing out of $T_\epsilon(M)$, and X be a vector field on M with isolated zeros. Then*

$$I(X) = \deg(G).$$

Proof. Recall that, in our previous notation, $T_\epsilon(M) = f_\epsilon(N_\epsilon)$ where $v \in \tau_x(M)^\perp$ and $f_\epsilon((x, v)) = x + v$. Then we define an extension $\tilde{X}$ of X to a vector field on $T_\epsilon(M)$ by

$$\tilde{X}(f_\epsilon((x, v)) = X(x) + v.$$

Since M is compact, for sufficiently small ϵ and $y \in S_\epsilon(M)$, $\tilde{X}(y)$ will point out of $T_\epsilon(M)$. So, by Theorem 7.4.5, $I(\tilde{X}) = \deg(G)$. Since $X(x) \in \tau_x(M)$ and $v \in \tau_x(M)^\perp$, we see that, for $y = f_\epsilon((x, v))$ in $T_\epsilon(M)$, $\tilde{X}(y) = 0$ if and only if both $v = 0$ and $X(x) = 0$: the singularities of $\tilde{X}$ are precisely those of X. Thus it remains to compare the indices of X and $\tilde{X}$ at such singularities.

By Proposition 7.3.6, we may assume that the singularities of X are non-degenerate and then we may use Proposition 7.3.5 to calculate the index at such a singularity p. Let ϕ be a chart on $\mathbb{R}^n$ at p, adapted to M with $\phi(p) = 0$

and let Φ be the corresponding chart on N_ϵ, as defined in the proof of Lemma 7.4.1 above. Then the first m elements of the standard basis of $\tau_{\phi(p)}(\mathbb{R}^n)$ are the images under $\phi_*(p)$ of a basis of $\tau_p(M)$ and we may replace the remainder by the images of the basis elements $e_{m+1}(p), \cdots, e_n(p)$ of $\tau_p(M)^\perp$, that we obtained in the proof of lemma. With respect to the resulting basis of $\mathbb{R}^n$, and equally of its tangent space at the origin, $\Phi_*(\tilde{X})$ will be given by a map $\tilde{F}$ with

$$D\tilde{F}(0) = \begin{pmatrix} DF(0) & 0 \\ 0 & I_{n-m} \end{pmatrix},$$

where F is the map expressing $\phi_*(X)$ in terms of these coordinates. Thus p is also a non-degenerate singularity of $\tilde{X}$ and $I(\tilde{X}, p) = I(X, p)$, so that $I(X) = I(\tilde{X}) = \deg(G)$, as required. ■

7.5. Morse Functions.

We next interpret the index of a vector field on a manifold in terms of invariants derived from suitably well-behaved functions on the manifold. The proof will require a function and a vector field that are closely related. However the fact that the index is independent of the vector field will imply that the same is true for the invariants derived from the appropriate functions.

Definitions 7.5.1. For a map $f : U \to \mathbb{R}$, where U is open in $\mathbb{R}^m$, a *critical point* is a point p for which $Df(p) = 0$, that is, $\frac{\partial f}{\partial x_i} = 0$ for $i = 1, \cdots, m$. A critical point is x *non-degenerate* if the *Hessian matrix* $Hf(x) = \left(\frac{\partial^2 f}{\partial x_i \partial x_j}(x)\right)$ is non-singular. Then a *Morse function* is one for which all the critical points are non-degenerate. Since the Hessian matrix is symmetric, it may be diagonalised by an orthogonal linear change of coordinates, which may be further re-scaled to give a diagonal matrix all of whose diagonal entries are ± 1. The number of entries that are -1 is called the *index of the* (non-degenerate) *critical point.*

Examples 7.5.2. 1) The function $f(x) = x^3$ on the real line has a degenerate critical point at the origin. The function $g(x) = \exp(-1/x^2)\sin^2(1/x)$ has the origin as a non-isolated critical point.

2) The function $f(x) = x^2$ on $\mathbb{R}^1$ has a non-degenerate critical point of index zero at the origin. However the function of two variables $f(x, y) = x^2$ has a non-isolated critical point, for which the index is therefore not defined, at every point of the x-axis, $y = 0$.

3) A critical point of index 1 of a function on $\mathbb{R}^2$ is often called a 'saddle point' because of the shape of its graph, illustrated in Figure 7.5(i). In Figure 7.5(ii) we illustrate the graph of the function $f(x, y) = x^3 - 3xy^2$ near the origin, a degenerate critical point often referred to as a 'monkey saddle'.

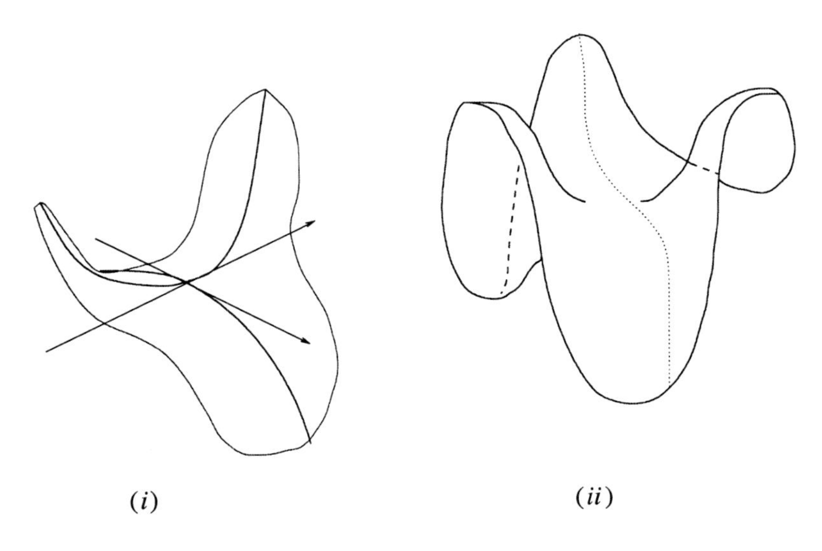

Figure 7.5.

Theorem 7.5.3. Every manifold M^m admits Morse functions.

Proof. We consider M^m embedded in some Euclidean space $\mathbb{R}^n$. Then, for each $v \in \mathbb{R}^n$ we can define the *height function*

$$h_v : M \longrightarrow \mathbb{R}^n; v \mapsto v \cdot x,$$

and almost all such height functions, that is all except those corresponding to the v that belong to a set of measure zero, are Morse functions.

To prove this first, for each point p in M, use an adapted chart at p to find a decomposition $\mathbb{R}^n = \mathbb{R}^m \times \mathbb{R}^k$ such that the projection of M onto the first factor is a diffeomorphism on some neighbourhood U_p of p. Since countably many such neighbourhoods cover M, it suffices to work in one of them.

Writing $v = (v_1, v_2) \in \mathbb{R}^m \times \mathbb{R}^k$ for $v \in \mathbb{R}^k$, let

$$f_v : \pi_1(U_p) \longrightarrow \mathbb{R}; x_1 \mapsto x \cdot v$$

be the coordinate representation of h_v. Let $g_v : \pi_1(U_p) \longrightarrow \mathbb{R}^m$ be the vector field grad f_v, see exercise 5.1(i), and note that the Jacobian matrix of g_v is the Hessian matrix of f_v. Then, for $w = (w_1, 0) \in \mathbb{R}^m \times \{0\}$, x_1 is a critical point of f_{v+w} if, and only if, $-w$ is a regular value of g_v. By Sard's Theorem this fails only for w in a set of measure zero. ■

We shall now show that non-degeneracy and the index are preserved by a change of coordinates and then, for a Morse function f on a manifold M, we let $\lambda_k(f)$ be the number of critical points of index k. Then our result is that

$$I(X) = \sum_{k=0}^{m} (-1)^k \lambda_k(f),$$

where X is a vector field so related to the given Morse function that its index is precisely the sum on the right hand side of this equation. To describe that relationship, we first prove the following result due to Morse.

Proposition 7.5.4. *Let p be a non-degenerate critical point of $f : U \to \mathbb{R}$ where $U \subset \mathbb{R}^m$. Then there is a local coordinate system $(y_1, \cdots, y_m)$ on a, possibly smaller, neighbourhood of p such that*

$$f(y_1, \cdots, y_m) = f(p) - y_1^2 - \cdots - y_k^2 + y_{k+1}^2 + \cdots + y_m^2$$

on this neighbourhood, where k is the index of the critical point p.

Proof. Without loss of generality we may assume both $p = 0$ and $f(p) = 0$ and then, as in the proof of Proposition 7.3.5, we may write

$$f(x_1, \cdots, x_m) = \sum_{j=1}^{m} x_j g_j(x_1, \cdots, x_m)$$

on some neighbourhood of the origin, where $g_j(x) = \int_0^1 \left(\frac{\partial f}{\partial x_j}(tx) \right) dt$. Then $g_j(0) = \frac{\partial f}{\partial x_j}(0) = 0$, since we are assuming p to be a critical point. Thus we may rewrite the functions g_j similarly to give

$$f(x_1, \cdots, x_m) = \sum_{i,j=1}^{m} x_i\, x_j\, h_{ij}(x_1, \cdots, x_m),$$

where the h_{ij} are smooth functions. Then $\frac{\partial^2 f}{\partial x_i \partial x_j}(0) = h_{ij}(0) + h_{ji}(0)$. So that, after symmetrising, the matrix $(h_{ij}(0))$ is $(1/2)Hf(0)$. We may now imitate the algebraic diagonalisation process for symmetric quadratic forms with our functions h_{ij} replacing the constants in the algebraic procedure. The only point we need to note is that the square root of a never zero smooth function is again a smooth function. Since $Hf(0)$ is assumed to be non-singular, we may make a linear change of coordinates to ensure that $h_{11}(0) \neq 0$ and define

$$y_1(x_1, \cdots, x_m) = \sqrt{|h_{11}(x_1, \cdots, x_m)|}\left(x_1 + \sum_{j>1} x_j h_{j1}(x_1, \cdots, x_m)\right).$$

Then $(y_1, x_2, \cdots, x_m)$ forms a new coordinate system on a neighbourhood of the origin for which

$$f = \pm y_1^2 + \sum_{i,j=2}^{m} x_i\, x_j\, k_{ij}(x_1, \cdots, x_m),$$

where $k_{ij}(0)$ is again a non-singular symmetric matrix, allowing the proof to continue by induction.

In terms of the y-coordinates $Hf(0)$ takes diagonal form with k entries equal to -2 and $m-k$ equal to $+2$. Since our coordinate transformation at the origin is non-singular, this number k must be the index of the original matrix $(1/2)Hf(0)$. ■

Corollary 7.5.5. *If V is another open subset of $\mathbb{R}^m$ and $\phi : V \to U$ is a diffeomorphism with $\phi(q) = p$, then $g = f \circ \phi$ has a non-degenerate critical point at q with the same index as that at p. Elsewhere g is non-critical.*

Proof. By the chain rule $Dg(x) = Df(\phi(x))\, D\phi(x)$ so, since $D\phi(x)$ is non-singular throughout V, $Dg(x) = 0$ if and only if $Df(\phi(x)) = 0$. If $\psi : W \to U$ is a change of coordinates such that $\psi(0) = p$ and $H(f \circ \psi)(0)$ is diagonal, then $\phi^{-1} \circ \psi$ maps 0 to q and transforms $Hg(0)$ to the identical diagonal form. Thus the indices agree at the two singularities. ■

This corollary allows us to transfer our concepts and results to oriented manifolds in the usual manner by working in terms of coordinate charts, since it says that they are unaffected by our choice of chart.

Definition 7.5.6. Given a non-degenerate critical point p of index k for a function f on M, we shall refer to a chart $\phi : U \to \mathbb{R}^m$ such that $\phi(p) = 0$ and

$$f \circ \phi^{-1}(y_1, \cdots, y_m) = f(p) - y_1^2 - \cdots - y_k^2 + y_{k+1}^2 + \cdots + y_m^2$$

as a *Morse chart* or *Morse coordinate system* for the critical point p. Then the Hessian matrix $H(f \circ \phi^{-1})(0)$ is diagonal with diagonal entries equal to ± 2.

The vector fields that are closely related to Morse functions, which we mentioned at the beginning of this section, are the so-called gradient-like fields:

Definition 7.5.7. Let $f : M \to \mathbb{R}$ be a Morse function. Then a vector field X on M is *gradient-like for f* if $Df(x) \cdot X(x) > 0$ whenever x is a non-critical point of f and if, when x is a critical point, there is a Morse chart $\phi : U \to \mathbb{R}^m$ for x such that $\phi_*(X|_U) = D(f \circ \phi^{-1})^T$.

Remark 7.5.8. Note that this definition ensures that the singular points of the gradient-like field X are precisely the critical points of the Morse function. $D(f \circ \phi^{-1})$ is a $(1 \times m)$-matrix and the vector $D(f \circ \phi^{-1})^T$ is usually known as $\operatorname{grad}(f \circ \phi^{-1})$. For a function g on $\mathbb{R}^m$, $\operatorname{grad} g(x)$, the *gradient of g at x* is characterised by the fact that, for all tangent vectors V at x, $Dg(x) \cdot V = \langle \operatorname{grad} g(x), V \rangle$. For vectors V of a given norm this is maximal when V is parallel to $\operatorname{grad} g(x)$. Thus we may think of $\operatorname{grad} g(x)$ as pointing in the direction of, and being proportional to, the maximum rate of change of g at the point x. Similarly, the gradient of a function may be defined intrinsically on a manifold using a Riemannian metric (see Definition 3.3.1) and it would be a gradient-like field in the above sense. However a change of metric would change the gradient field of a given function.

Proposition 7.5.9. *Every Morse function on a manifold admits a gradient-like vector field.*

Proof. Note that Proposition 7.5.4 implies that non-degenerate critical points must be isolated, so we may choose an atlas $\mathcal{U} = \{(U_\alpha, \phi_\alpha) \mid \alpha \in A\}$ for M such that each coordinate neighbourhood U_α contains at most one critical point of f and, when it does, ϕ_α is a Morse chart for that critical point. We may also assume, in the latter case, that there is a neighbourhood of the critical point that meets no other coordinate neighbourhood. This is trivially possible when M is compact. However it is also possible under the hypothesis of paracompactness. On U_α we define

$$X_\alpha = (\phi_\alpha^{-1})_* \left(\operatorname{grad}(f \circ \phi_\alpha^{-1})\right).$$

Then, if $\{\rho_\alpha \mid \alpha \in A\}$ is a partition of unity subordinate to $\mathcal{U}$, we may, as usual, extend $\rho_\alpha X_\alpha$ to a vector field on M by defining it to be zero outside of U_α and then define $X = \sum_{\alpha \in A} \rho_\alpha X_\alpha$. This means that, if x is a critical point of f, it lies in a unique U_α and $\rho_\alpha = 1$ on some neighbourhood V_α of x in U_α. Then $X = X_\alpha$ on V_α and $\phi_\alpha\big|_{V_\alpha}$ is a Morse chart for x. Moreover $\left(\phi_\alpha\big|_{V_\alpha}\right)_* \left(X\big|_{V_\alpha}\right) = \mathrm{grad}(f \circ {\phi_\alpha}^{-1})$, as required. If x is not critical then, by the linearity of $Df(x)$, $Df(x) \cdot (X(x)) = \sum_\alpha \rho_\alpha(x)\,(Df(x) \cdot X_\alpha(x))$. This is strictly positive since, for at least one α, $\rho_\alpha(x) > 0$ and, whenever $x \in U_\alpha$,

$$\begin{aligned} Df(x) \cdot X_\alpha(x) &= D(f \circ \phi_\alpha^{-1})(\phi_\alpha(x)) \cdot \big(\mathrm{grad}(f \circ \phi_\alpha^{-1})(\phi_\alpha(x)\big) \\ &= \|\,\mathrm{grad}(f \circ \phi_\alpha^{-1})(\phi_\alpha(x))\,\|^{\,2}, \end{aligned}$$

which is stricly positive since x is not a critical point. Thus X is indeed gradient-like for the Morse function f. ■

We are now ready to establish the main result of this section.

Theorem 7.5.10. *Let X be a vector field with isolated singularities on a compact manifold M of dimension m and f be a Morse function on M with λ_k critical points of index k. Then*

$$I(X) = \sum_{k=0}^{m} (-1)^k \lambda_k.$$

Proof. We have already seen that all vector fields on M have the same index, so we may assume that X is a gradient-like field for f. Then its critical points are the singular points of f and the result will follow from the fact that, if p is a critical point of f of index k, then $I(X, p) = (-1)^k$. To see that let (U, ϕ) be a Morse chart for f at p, so that $g = f \circ \phi^{-1}$ takes the form

$$g(y_1, \cdots, y_m) = f(p) - y_1^2 - \cdots - y_k^2 + y_{k+1}^{\,2} + \cdots + y_m^{\,2}$$

and $I(X,\, p) = I(\mathrm{grad}\, g,\, 0)$, since $\phi_*\left(X\big|_U\right) = \mathrm{grad}\, g$. However

$$\mathrm{grad}\, g(y_1, \cdots, y_m) = 2(-y_1, \cdots, -y_k, y_{k+1}, \cdots, y_m)$$

so that $\mathrm{grad}\, g$, when interpreted as a diffeomorphism as in Section 7.2, is the linear map with matrix the Hessian matrix $Hg(0)$. This has determinant $(-2)^k$ so, by Proposition 7.3.4, $I(\mathrm{grad}\, g,\, 0) = (-1)^k$. ■

This theorem has a rather remarkable consequence.

Corollary 7.5.11. *If M is a compact manifold of odd dimension, then any vector field on M with isolated singularities has index zero.*

Proof. If f is a Morse function on M, then so too is $-f$ defined by $(-f)(x) = -(f(x))$. The critical points of $-f$ are the same as those of f. However, if x is a critical point of f with index k, then as a critical point of $-f$ it has index $m-k$. Since $(-1)^{m-k} = -(-1)^k$, the result follows from the identity in the theorem. ■

7.6. The Euler Number.

If M^m is a compact manifold, then the space $H^k(M^m)$ is finite dimensional and its dimension is called the *kth Betti number* $\beta_k(M)$ *of* M. Then the *Euler number* $\chi(M)$ *of* M is defined to be the alternating sum $\sum_{k=0}^{m}(-1)^k\beta_k$. Thus

$$\chi(\mathbf{S}^{2n+1}) = 0, \quad \chi(\mathbf{S}^{2n}) = 2, \quad \chi(T^n) = \sum_{k=0}^{n}(-1)^k\binom{n}{k} = (1-1)^n = 0.$$

First we note the following consequence of the Mayer-Vietoris sequence.

Proposition 7.6.1. *Let U and V be open subsets of a manifold such that U, V and $U \cap V$ all have finite dimensional de Rham cohomology spaces. Then so does $U \cup V$ and*

$$\chi(U \cup V) = \chi(U) + \chi(V) - \chi(U \cap V).$$

Proof. This follows from the long exact Mayer-Vietoris sequence:

$$\longrightarrow H^{p-1}(U \cap V) \longrightarrow H^p(U \cup V) \longrightarrow H^p(U) \oplus H^p(V) \longrightarrow H^p(U \cap V) \longrightarrow$$

This shows firstly that $H^p(U \cup V)$ is finite dimensional for each p. Then the result follows from the fact that the alternating sum of the dimensions of the vector spaces in an exact sequence is zero. ■

Example 7.6.2. Using this result the Euler number is often easier to calculate directly rather than via the Betti numbers. Thus the fact that the sphere is the union of two open discs (each having Euler number 1) whose intersection is homotopy equivalent to a sphere of one lower dimension, and the fact that the zero sphere is two points (so Euler number 2) means that the Euler numbers of spheres are alternately two and zero as stated above.

The main result in this section will be the Poincaré-Hopf Theorem that, for any vector field X on M with isolated singularities, $\chi(M) = I(X)$. We shall prove this using the relation between $I(X)$ and Morse functions that was proved in the previous section and we shall also require two further lemmas. These both concern the manner in which a non-degenerate function determines the global structure of a manifold and go some way toward explaining the mystery of all vector fields on a manifold having the same index, provided only that their singularities be non-degenerate. In order to state these lemmas, given a Morse function f on a manifold M, we shall denote by $M(a, b)$ the set $\{x \in M \mid a \leqslant f(x) \leqslant b\}$ and by $M(a)$ the level hypersurface $\{x \in M \mid f(x) = a\}$.

Lemma 7.6.3. *Let $a < b < c \in \mathbb{R}$ and $M(b, c)$ be compact and contain no critical points. Then $M(a, b)$ is diffeomorphic with $M(a, c)$.*

Proof. Choose a gradient-like field X for f. Then, for any integral curve $\gamma(t)$ of X, we have

$$(f \circ \gamma)'(t) = Df(\gamma(t)) \cdot \gamma'(t) = Df(\gamma(t)) \cdot X(\gamma(t)) > 0,$$

whenever $\gamma(t) \in [b, c]$. Thus the function f is increasing along γ and, since $M(b,c)$ is compact, it follows that any integral curve γ_x of X that starts at $x \in M(b)$ at 'time' $t = 0$ will reach $M(c)$ at a time $\tau(x)$ where τ is a smooth function on $M(b)$. Let $\lambda : M \to \mathbb{R}$ be a smoth function equal to $1/\tau$ on $M(b)$, constant along the integral curves from $M(b)$ to $M(c)$ and vanishing outside some compact neighbourhood K of $M(b,c)$ in M. Then the vector field $Y(x) = \lambda(x)X(x)$ admits a 1-parameter group of diffeomorphisms ρ_t by Lemma 2.7.5. Since the integral curves of Y that start on $M(b)$ at $t = 0$ reach $M(c)$ at $t = 1$, it follows that ρ_1 is the diffeomorphism we seek. ■

For our present purpose this lemma implies, of course, that, if the Euler number $\chi(M(a,b))$ is defined, so too is $\chi(M(a,c))$ and is equal to it. When $M(b,c)$ does contain critical points, then the Euler number will change. However the difference may be calculated, once again using a gradient-like vector field. Since critical points are isolated and, in the following proof, we may work on an arbitrarily small Morse neighbourhood of each critical point, it suffices to consider the case that there is only one critical point in $M(b,c)$.

Lemma 7.6.4. *Let $a < b < c < d \in \mathbb{R}$ and $M(b,d)$ be compact. Assume that $M(b,d)$ contains exactly one critical point x, that x has index k and $f(x) =$*

c. Then, if $\chi(M(a, b))$ *is defined, so too is* $\chi(M(a,d))$ *and* $\chi(M(a,d)) = \chi(M(a,b)) + (-1)^k$.

Proof. Choose a gradient-like field X for f on $M(a,d)$ and work on the coordinate space of a Morse chart (ϕ, U) for the critical point, defining $g = f \circ \phi^{-1}$ on $\phi(U)$. We may find $\delta > 0$ such that the ball $B(0, \delta)$ is contained in $\phi(U)$ and then, since $g(0) = c$, choosing $0 < \epsilon < \delta^2$ will ensure that both $g^{-1}(c - \epsilon)$ and $g^{-1}(c + \epsilon)$ are non-empty. Using the previous lemma, we may also assume that $b = c - \epsilon$ and $d = c + \epsilon$. Writing $\mathbb{R}^m = \mathbb{R}^k \times \mathbb{R}^{m-k}$ with coordinate vectors x and y in the respective factors, we have $g(x, y) = c - ||x||^2 + ||y||^2$. The situation is illustrated in Figure 7.6. Note that the level hypersurfaces are only disconnected because we have necessarily illustrated the unrepresentative case $k = m - k = 1$ and $\mathbf{S}^0$ comprises two points. We have taken advantage of this to illustrate different aspects of our proof on different parts of the diagram.

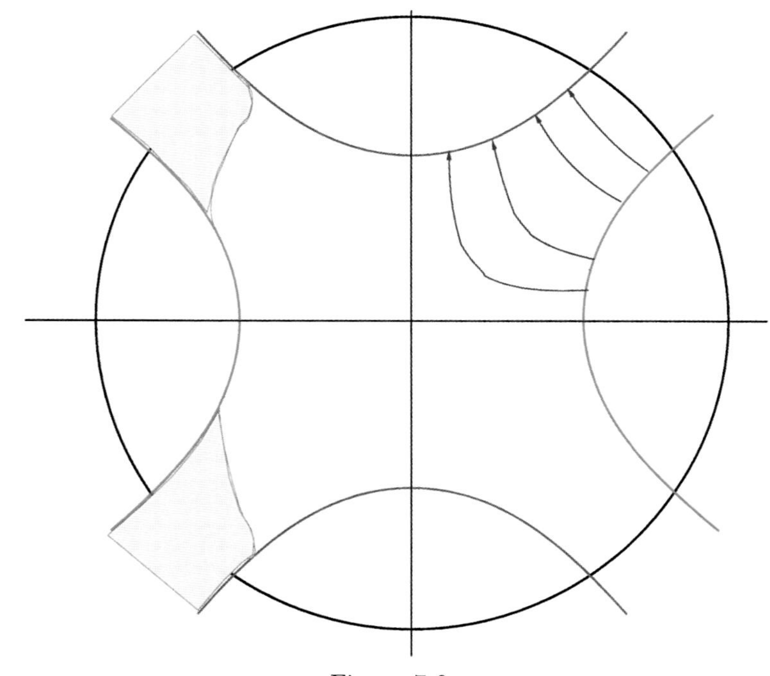

Figure 7.6.

Since we have explicit coordinates, we note at once that the intersection

$$B = \phi(M(b, d)) \cap B(0, \delta)$$

is diffeomorphic with an m-dimensional ball. The gradient-like field for f is mapped by ϕ_* to the actual gradient field for g on the part of the coordinate space illustrated. In the first quadrant we indicate (in blue) the integral curves of this field that lie within $M(b,d)$. The intersection of the (red) $(c-\epsilon)$-level with the (black) sphere of radius δ is the product $\mathbf{S}^{k-1} \times \mathbf{S}^{m-k-1}$ of the spheres given, respectively, by $\|x\|^2 = (\delta^2+\epsilon)/2$ and $\|y\|^2 = (\delta^2-\epsilon)/2$. Similarly the intersection of the sphere with the (green) $(c+\epsilon)$-level is the diffeomorphic product of the spheres given, respectively, by $\|x\|^2 = (\delta^2-\epsilon)/2$ and $\|y\|^2 = (\delta^2+\epsilon)/2$. We consider a product neighbourhood in the $(c-\epsilon)$-level of its intersection with the δ-sphere that is diffeomorphic with $\mathbf{S}^{k-1} \times \mathbf{S}^{m-k-1} \times (-1,\,1)$, where $\mathbf{S}^{k-1} \times \mathbf{S}^{m-k-1} \times \{0\}$ corresponds to the intersection and $\mathbf{S}^{k-1} \times \mathbf{S}^{m-k-1} \times (-1,\,0)$ lies inside $B(0,\delta)$. Choosing a' sufficiently close to b that $M(a',b)$ contains no critical points of f and using a smooth function λ on $(-1,\,1)$ that is 0 on $(-1,-3/4]$ and 1 on $[-1/4,\,0]$, we may, as in the proof of the previous lemma, produce a modification Y of the gradient-like field X such that each integral curve of Y starting at time $t=0$ on $M(a')$ arrives at time $t=1$ on

(*i*) $M(d)$ along curves that do not meet $\phi^{-1}(B(0,\delta))$;

(*ii*) $M(b)$ along curves that meet $\phi^{-1}(B(0,\delta))$ but whose images in $B(0,\delta)$ do not meet the product neighbourhood $\mathbf{S}^{k-1} \times \mathbf{S}^{m-k-1} \times (-1,\,1)$;

(*iii*) a fraction $\lambda(t)$ of the way along the integral curve from $M(b)$ to $M(d)$ along curves whose images in $B(0,\delta)$ meet $\mathbf{S}^{k-1} \times \mathbf{S}^{m-k-1} \times \{t\}$.

Then the diffeomorphism ρ_1 from the associated 1-parameter group of diffeomorphisms will map $M(a,b)$ onto the modified version $\widetilde{M}(a,b)$. The additional part $\widetilde{M}(a,b) \setminus M(a,b)$ is indicated, in magenta, on the left-hand side of the diagram. Although $\widetilde{M}(a,b)$ is not an open subset of $M(a,d)$ we can make it so by removing that part of its (upper) boundary whose image lies in the open ball $B(0,\delta)$. That done, we note, again from the explicit nature of the construction, that the intersection of $\widetilde{M}(a,b)$ with the ball $B = \phi(M(b,d)) \cap B(0,\delta)$ is diffeomorphic with a product $\mathbf{S}^{k-1} \times B^{m-k+1}$, where B^m denotes the unit ball in $\mathbb{R}^m$. It follows from Proposition 7.6.1 that

$$\begin{aligned}
\chi(M(a,d)) &= \chi(B) + \chi(\widetilde{M}(a,b)) - \chi(\mathbf{S}^{k-1}) \\
&= 1 + \chi(M(a,b)) - (1 + (-1)^{k-1}) \\
&= \chi(M(a,b)) + (-1)^k.
\end{aligned}$$

Here we have used the facts that $\mathbf{S}^{k-1} \times B^{m-k+1}$ is homotopy equivalent to $\mathbf{S}^{k-1}$, that $\widetilde{M}(a,b)$, even with part of its boundary removed, is still homotopy equivalent to $M(a, b)$ and that homotopy equivalent manifolds have the same cohomology and hence the same Euler numbers. ∎

Theorem 7.6.5. (Poincaré-Hopf.) *If M is a compact, oriented manifold without boundary and X is a vector field on M with only isolated zeros, then $\chi(M) = I(X)$.*

Proof. Since Morse functions exist on M by Theorem 7.5.3, this follows from the preceding lemmas. Note that, since M is compact, $M(a)$ will be empty for some $a \in \mathbb{R}$. Then a point at which a Morse function takes its minimum value will be a critical point of index 0 contributing a summand $+1$ to the Euler number. We may separate the critical levels, those containing critical points, by non-critical ones and then apply Lemma 7.6.4 to the resulting 'slices' of the manifold M. If two or more critical points take the same value we may modify the Morse function, in a manner equivalent to the manipulation of handles that we describe in the next section, to ensure that there is only one critical point in each slice as required for the Lemma. ∎

Corollary 7.6.6. *If M^m is an oriented compact connected closed hyper-surface in $\mathbb{R}^{m+1}$, then $\chi(M) = \deg(G)$, where G is the Gauss map associated with the positive unit normal field on M.* ∎

Corollary 7.6.7. *If M is a compact, oriented manifold without boundary such that $\chi(M) \neq 0$, then M does not admit a never-zero vector field.*

Proof. A vector field with no zeros must have index zero. ∎

This establishes the result we mentioned in Chapter 3: since $\chi(\mathbf{S}^2) = 2$, any vector field on the 2-sphere has a zero. Thinking of the vector at a point as the projection on the tangent space of a hair growing out of the sphere at that point, this is popularly interpreted as the statement that a hairy ball cannot be combed flat. Of course the same is true for any even dimensional sphere.

In fact, although it goes somewhat beyond the scope of this book to prove it, the Euler characteristic is the precise obstruction to the existence of a never-zero vector field: $\chi(M) = 0$ if, and only if M admits a smooth field of non-zero tangent vectors. Necessity may be established explicitly in special cases – we have already done it for $\mathbf{S}^1$ – but for a general proof see Steenrod (1951) or Husemoller (1991).

7.7. Handle Decompositions.

In this final section of this chapter we show how our analysis of the neighbourhood of a critical point of a Morse function leads to a particularly useful decomposition of the manifold and outline some important applications. We leave it as an extended exercise for the reader to fill in the details consulting, if necessary, the relevant texts that we mention in our survey of the literature, as well as Milnor (1965[1])

7.7.1. Existence of handle decompositions.

Lemma 7.6.4 or, more precisely, its proof has more to it than we used there. Referring to the details in that proof, we showed that the part $M(a, b)$ of the manifold between the 'levels' a and b of the Morse function is diffeomorphic with a modified version $\widetilde{M}(a, b)$ and then $M(a, d)$, which includes the critical point at level c, is $\widetilde{M}(a, b)$ together with an overlapping ball B. If, instead of B, we look at $\widetilde{B} = B \setminus \widetilde{M}(a, b)$, then $\widetilde{B}$ is homeomorphic, and almost diffeomorphic, to B and it is not difficult to see that it meets $\widetilde{M}(a, b)$ along a submanifold diffeomorphic with $\mathbf{S}^{k-1} \times B^{m-k}$ and that it meets the d-level, the upper boundary of $M(a, d)$, along a submanifold diffeomorphic with $B^k \times \mathbf{S}^{m-k-1}$. The diffeomorphism of $\widetilde{B}$ with B only fails along the intersection, $\mathbf{S}^{k-1} \times \mathbf{S}^{m-k-1}$, of these two submanifolds where our construction of $\widetilde{M}(a, b)$ introduced a submanifold of cusps. Thus we have, on passing a single critical point of index k,

$$M(a, d) \cong M(a, b) \cup B^k \times B^{m-k}$$

where the ball $B^k \times B^{m-k}$ is 'attached' by a diffeomorphism of $\mathbf{S}^{k-1} \times B^{m-k}$ with a submanifold of $M(b)$ and cusps are introduced along $\mathbf{S}^{k-1} \times \mathbf{S}^{m-k-1}$ in order to make the result a smooth manifold. It can be shown that the introduction of cusps is canonical: the resulting manifold is unique up to diffeomorphism (Cerf, 1961). In honour of the special case $k = 1$, $m = 3$, this procedure is referred to as attaching a handle and the resulting presentation of the entire manifold, starting either from the empty set or from a union of boundary components, is referred to as a *handle decomposition.*

Handle decompositions give a good insight into the structure and properties of manifolds. The most dramatic application is Smale's proof (1961) that a differential manifold of dimension at least five that is homotopy equivalent to a sphere is homeomorphic to that sphere. (See, however example 3.4.6) Unfortunately, although handle decompositions still exist in lower dimensions, the techniques for manipulating them do not: the analogous result in dimension three, for which Poincaré originally conjectured it, still remains open.

It is, for example, worth noting that the procedure by which we obtained a handle decomposition from a gradient-like field for a Morse function is reversible. Given a handle decomposition from any source, it is clear from Figure 7.6 that we could inscribe on each k-handle a vector field with index k that was everywhere transverse (that is, never tangent) to the boundaries of that handle. Then, making a smooth adjustment to line up the resulting vector fields across the boundaries of the handles, we obtain a gradient-like vector field X on M, though not immediately a function to which it relates. Then, reinterpreting Lemma 7.6.4 with $M(b, d)$ being the given handle, it follows that $I(X) = \chi(M)$. Thus we have shown:

Theorem 7.7.1. *If the compact manifold-without-boundary M^m has a handle decomposition with a total of γ_k k-handles, then $\chi(M) = \sum_0^m (-1)^k \gamma_k$.* ■

Corollary 7.7.2. *If the dimension m is odd, then $\chi(M) = 0$.*

Proof. Replacing the Morse function f that gave rise to the handle decomposition by $-f$, in effect turns the handle decomposition upside down, with each k-handle becoming an $(m-k)$-handle. Thus $\chi(M) = \sum_0^m (-1)^{m-k} \gamma_k = -\chi(M)$. ■

Of course, this is just another way of looking at the proof of Corollary 7.5.11.

7.7.2. Manipulating handle decompositions.

The value of a handle decomposition stems, not only from its relative simplicity and explicit nature, but also from the possibility of manipulating and rearranging the handles themselves as was mentioned briefly above and will be explained in more detail at the end of the next subsection. The appropriate manner of manipulation will be through *isotopies*, that is, homotopies of the identity map of the manifold, each stage of which is a diffeomorphism. If we re-examine Figure 7.6 and its description, we see that $M(b, d)$ is diffeomorphic to a product of the lower boundary with the unit interval, once we remove the 'core' $B^k \times \{0\}$, represented by the segment of the horizontal black line between the red curves in Figure 7.6, and the 'co-core' $\{0\} \times B^{m-k}$, represented by the segment of the vertical black line between the green curves, of the k-handle. When a k-handle is attached immediately after an ℓ-handle where $k \leqslant \ell$, we note that the 'attaching' $(k-1)$-sphere, the boundary of the core, of the k-handle and the 'belt' $(m-\ell-1)$-sphere, the boundary of the co-core of the ℓ-handle, both lie in a common level surface and have total dimension less than that of the surface. Using bump functions and the fact, a corollary of transversality theorems, c.f. Bröcker and Jänich (1982), that a small isotopy of a manifold M^m will move a submanifold N^n so that is does not meet another, fixed, submanifold P^p provided that $n+p<m$, that is, the dimension of N is less than the codimension of P, it is then not difficult to construct an isotopy of the manifold, fixed away from the two handles, that produces a new decomposition in which the k-handle is attached first.

Repeating this process as often as necessary, we arrive at a handle decomposition in which the necessary 0-handles are added first, then the 1-handles, and so on. Indeed we may also arrange that, for each k, all the k-handles are attached independently and simultaneously.

Another important manipulation involves two handles of the same type: the procedure referred to as 'adding one handle to another'. After attaching a first k-handle the upper boundary contains a copy of the attaching sphere bounding a k-disc. We may use this disc to produce an isotopy of the attaching sphere of a second k-handle to a position where it corresponds to the connected sum of the two original attaching spheres.

7.7.3. Cohomology via handle decompositions.

With a little more effort we can be more precise about Theorem 7.7.1. If a manifold M_2 is expressed as M_1 with a k-handle attached we can, referring to the description above, replace the precise handle $B^k \times B^{m-k}$ by a fatter version $\mathcal{H}^k$ which is the original extended by a 'collar', a neighbourhood in $M(a,\, b)$ that is the product with the half open interval $[0,\, 1)$ of the image in $M(b)$ of the diffeomorphism that attaches the handle. Then we have expressed M_2 as the union of two open subsets M_1 and $\mathcal{H}^k$ where $M_1 \cap \mathcal{H}^k \cong \mathbf{S}^{k-1} \times B^{m-k+1} \simeq \mathbf{S}^{k-1}$. Applying the Mayer-Vietoris theorem to this decomposition we have, since $\mathcal{H}^k$ is contractible, the exact sequence

$$0 \longrightarrow H^{k-1}(M_2) \longrightarrow H^{k-1}(M_1) \longrightarrow \mathbb{R} \xrightarrow{\delta} H^k(M_2) \longrightarrow H^k(M_1) \longrightarrow 0. \tag{7.1}$$

Here $\mathbb{R}$ is $H^{k-1}(\mathbf{S}^{k-1})$ so we require $k > 1$; the first zero is $H^{k-2}(\mathbf{S}^{k-1})$ for $k > 2$ or is deducible from the previous homorphism being surjective when $k = 2$; the final zero is $H^k(\mathbf{S}^{k-1})$. Since this a sequence of vector spaces, the coboundary homomorphism δ is either injective or zero. In the former case it adds a new summand $\mathbb{R}$ to $H^k(M_1)$ to form $H^k(M_2)$, that is, it increases its dimension by 1, and in the latter case it removes such a summand from $H^{k-1}(M_1)$ to form $H^{k-1}(M_2)$ so decreasing its dimension by 1.

A handle decomposition of a manifold M might start from some boundary component if it has one. However we are usually interested in a full handle decomposition starting from the empty set. Then we must start with a 0-handle, that is an n-ball, 'attached' by the 'empty map' to the empty set: in other words we simply introduce an n-ball whose boundary sphere may be used to attach further handles that have non-empty attaching maps. Using the manipulations of the previous subsection we may asssume, for each k, that all the k-handles are added simultaneously and before the $(k+1)$-handles. The 1-handles are trivial to analyse: the attaching $\mathbf{S}^0 \times B^{n-1}$ either joins two previous components, thus reducing the dimension of H^0 or has both ends on the same component in which case it increases the dimension of H^1. In the former case, since only 0-handles preceded the 1-handle, we could omit both the 1-handle and one of the 0-handles from the decomposition.

Thereafter any summand of H^{k-1} 'killed' by a k-handle would have to be a linear combination of those that have been added by preceding $(k-1)$-handles: those, that is, that did not kill summands introduced by $(k-2)$-handles. The essence of Smale's proof of the Poincaré conjecture is to perform a sequence of additions of the $(k-1)$-handles to each other to produce a new one that represents the linear combination of the old ones that is 'algebraically cancelled' by a k-handle. He then further manipulates such algebraically cancelling pairs of handles to replace them by ones that cancel geometrically and so may be removed from the decomposition altogether. This is the step, analogous to the cancellation of 0-handles and 1-handles above, that fails in the lower dimensions. Since the cohomology of a homotopy sphere, like that of the sphere itself, only requires one 0-handle and one n-handle to realise it, and since it is possible to prove that any manifold with such a handle decomposition is indeed a sphere, the result follows. We should emphasize however that the cohomology, or more directly the homology, involved in Smale's proof is that with integer coefficients, rather than the de Rham version that we are using. Nevertheless it still only requires the same two handles to express the integral cohomology of a homotopy sphere.

7.7.4. Identification with other cohomology groups.

In algebraic topology one usually defines homology, and so also cohomology, theories with coefficients drawn from an arbitrarily chosen abelian group, if not from an even more general source. However if we restrict such theories to have the real numbers for their coefficients then we recover precisely the de Rham cohomology, and dual homology, spaces. The usual way to identify two apparently different cohomology theories is to check that they satisfy the Eilenberg-Steenrod axioms (Eilenberg and Steenrod, 1952). We have in fact established several of these axioms for de Rham cohomology: (1) that it is a Functor, (smooth) maps f between manifolds induce homomorphisms f^* between the cohomology spaces such that $\mathrm{id}^* = \mathrm{id}$ and $(fg)^* = g^* f^*$; (2) the Homotopy Axiom, that homotopic maps induce the same homomorphisms of the cohomology spaces; (3) the Dimension Axiom, that $H^k(\mathrm{pt}) = 0$ for $k \neq 0$, a consequence of the axioms being that $H^0(\mathrm{pt})$ becomes the coefficients of the theory.

The remaining axioms all involve the the cohomology of a pair of spaces which, for most theories, would be rather general. However for our purposes we may restrict attention to pairs (M, N) where N is a, possibly empty, submanifold of M and, replacing it if necessary by a tubular neighbourhood which, by the homotopy axiom, would not affect the cohomology, we may further assume that N is a submanifold-with-boundary of the same dimension as M. We define $\Omega^k(M, N)$ to be the space of k-forms on M that are zero on N. Since the same is true of the exterior derivative of such a form we have a chain complex as before that leads to the cohomology groups $H^k(M, N)$. Note that $H^k(M, \emptyset)$ is precisely our previous $H^k(M)$. Any map of pairs $f : (M, N) \longrightarrow (M', N')$, where $f(N) \subseteq N'$, clearly induces a pull-back homomorphism $f^* : H^k(M', N') \longrightarrow H^k(M, N)$ which, once again, satisfies the homotopy axiom. Moreover it is not difficult to see that the inclusions of pairs

$$N \xrightarrow{i} M \xrightarrow{j} (M, N)$$

produce a ladder of horizontal short exact sequences

$$0 \longrightarrow \Omega^k(M, N) \xrightarrow{j*} \Omega^k(M) \xrightarrow{i*} \Omega^k(N) \longrightarrow 0$$

forming vertical chain complexes, similar to those that gave rise to the Mayer-Vietoris exact sequence. As in that case this ladder also gives rise to a coboundary homomorphism $\delta : H^k(N) \longrightarrow H^{k+1}(M, N)$ which commutes with the 'pull-back' homomorphisms and fits in a long exact sequence

$$\cdots \longrightarrow H^k(M,\, N) \xrightarrow{j*} H^k(M) \xrightarrow{i*} H^k(N) \xrightarrow{\delta} H^k(M,\, N) \longrightarrow \cdots.$$

The final axiom, the Excision axiom, states that if P is a submanifold-with-boundary contained in the interior of N then inclusion induces an isomorphism $H^k(M,\, N) \cong H^k(M \setminus \mathrm{int}P,\, N \setminus \mathrm{int}P)$, which is immediate from our revised definition.

Thus, since de Rham cohomology satisfies all the Eilenberg-Steenrod axioms it coincides with any other cohomology theory that satisfies them and has $H^0(\text{point}) \cong \mathbb{R}$.

In fact from these additional axioms may be deduced the Mayer-Vietoris sequence that we had already proved for de Rham cohomology and we may note that the first three axioms, together with the Mayer-Vietoris sequence, were all that we used in computing the cohomology spaces of the spheres and then in establishing the exact sequence (7.1). Thus to show directly that another cohomology theory with real coefficients is the same as the de Rham theory it suffices to show that the coboundary map δ is the same in both cases. It turns out this is determined by the degree of the map of the attaching sphere of the k-handle on each of the $(k-1)$-handles that introduce generators of H^{k-1}. That in turn may be determined by using the product structure, away from the cores and cocores of the handles, to extend the attaching sphere of a k-handle until it meets the belt spheres of the $(k-1)$-handles transversely in isolated points. The sum of the intersection numbers (in terms of appropriate induced orientations) at these points for each relevant $(k-1)$-handle is the required degree.

7.8. Exercises for Chapter 7.

7.1. Prove that the antipodal map $\alpha(p) = -p$ on the unit sphere in $\mathbb{R}^m$ has degree $(-1)^m$.

Deduce that the real projective space $\mathbb{RP}^n$ is not orientable when n is even.

7.2. Identifying $\mathbf{S}^1$ with $\{z \in \mathbb{C} \,|\, |z| = 1\}$ find the degree of the map $\theta_n : z \mapsto z^n$.

Let $\theta : [0,1] \to \mathbb{R}$ be such that $\theta(0) = 0$ and $\theta(1) = 2\pi n$ and let $r : [0,1] \to (0,\infty)$ be such that $r(0) = r(1)$. Find the winding number about a of the curve

$$\gamma(t) = a + r(t)\ \exp(i\theta(t)).$$

7.3. The linking number of two disjoint simple closed curves in $\mathbf{S}^3$ may be defined by removing a point through which neither curve passes and transferring the definition in $\mathbb{R}^3$ via a diffeomorphism with $\mathbf{S}^3 \setminus \{p\}$. The Hopf map $\mathbf{S}^3 \to \mathbf{S}^2$ may be defined by identifying $\mathbf{S}^3$ with $\{(z_1, z_2) \in \mathbb{C}^2 \mid |z_1|^2 + |z_2|^2 = 1\}$ and mapping

$$(z_1, z_2) \mapsto z_1/z_2 \in \mathbb{C} \cup \{\infty\} \cong \mathbf{S}^2.$$

Show that every point of $\mathbf{S}^2$ is regular and find the linking number of the inverse images of any two points.

7.4. Construct a map $f : \mathbf{S}^2 \to \mathbf{S}^2$ of degree n for each $n \in \mathbb{Z}$. Do the same for all spheres $\mathbf{S}^m$.

7.5. Check the result of Theorem 7.6.5 for various examples on $\mathbf{S}^1$ and $\mathbf{S}^2$.

7.6. Find a never zero vector field on the spheres $\mathbf{S}^{2k-1}$.

CHAPTER 8

LIE GROUPS

In his assessment of the work of Elie Cartan that was published in 1914, Henri Poincaré wrote:

> *Toute théorie mathématique est, en dernière analyse, l'étude des propriétés d'un groupe d'opérations,*

and continued by describing Lie groups as

> *ceux auxquels se rattachent les principales théories géométriques,*

to which he might have added *'et la théorie des particules élémentaires en physique'.* Our aim in this chapter is to provide a very brief introduction to this very important subject, restricting ourselves to little more than definitions and motivational examples. The beauty and the power of the subject are rooted in the way in which it combines the differential topology, which we have been studying, with abstract group theory and analysis. In a number of exercises and examples we have already surreptitiously introduced these groups, which serve as marvelous illustrations in the theory of submanifolds, fibre bundles and de Rham cohomology. The interested reader is urged to turn to some systematic text, such as Chevalley (1946), Chevalley and Eilenberg (1948) or Helgason (1978) for a more detailed presentation.

8.1. Lie Groups.

Definition. 8.1.1. A *Lie group* is a group G equipped with a smooth manifold structure, such that the group operations $(x, y) \mapsto xy$ and $x \mapsto x^{-1}$, $x, y \in G$, are smooth maps.

A homomorphism $\varphi : G_1 \to G_2$ of Lie groups is a map that is both compatible with the group structures and smooth as a map of manifolds.

It is a deep, but not very useful, result that every connected, locally Euclidean $\mathcal{C}^\circ$-group has a compatible $\mathcal{C}^\infty$-structure. The question was posed by D. Hilbert in 1900, and solved by A. Gleason, D. Montgomery and L. Zippin in 1952. Granted this, it is not hard to show that the $\mathcal{C}^\infty$-structure is unique: this follows from a theorem that we shall sketch below concerning closed subgroups of Lie groups.

Let x be an arbitrary element of the Lie group G. The symbol L_x denotes left translation in G by x, that is, $L_x(y) = xy$ for all $y \in G$. The defining properties of G imply that L_x is a diffeomorphism, with inverse $L_{x^{-1}}$, and the derivative $(L_x)_* : \tau(G) \to \tau(G)$ must also be a diffeomorphism. For each tangent space $\tau_y(G)$, $(L_x)_* : \tau_y(G) \to \tau_{xy}(G)$ is a linear isomorphism.

Definition 8.1.2. The vector field X on G is said to be *left-invariant* if and only if

$$(L_x)_* X(y) = X(xy), \qquad \text{for all points } x, y \in G,$$

or equivalently, $(L_x)_* \circ X = X \circ L_x$.

Such a field X is completely determined by its value at the identity $X(e)$. Conversely for every vector $v \in \tau_e(G)$ we can define a (smooth) left-invariant vector field X on G by $X(y) = (L_y)_*(v)$. This proves

Lemma 8.1.3. *The map $X \mapsto X(e)$ is a linear isomorphism between the set of left-invariant vector fields on G and $\tau_e(G)$, the tangent space at the identity.* ∎

Corollary 8.1.4. *If G is a Lie group, then the tangent bundle $\tau(G)$ is trivial.*

Proof. Let $\{v_1, \cdots, v_n\}$ be a basis for the vector space $\tau_e(G)$ and write $X_i(y) = (L_y)_*(v_i)$ for all $y \in G$, $i = 1, \cdots, n$. Then $\{X_1, \cdots, X_n\}$ is a basis of sections for $\tau(G)$, and provides a trivialisation by Theorem 3.2.3. ∎

Definition 8.1.5. A (Lie) subgroup of a Lie group G is an injective homomorphism of Lie groups $i : H \to G$.

Although i is always an *immersion*, the following example shows that it need not be an *embedding* of submanifolds.

Example 8.1.6. Define $\varphi : \mathbb{R} \to T^2$ by

$$\varphi(t) = (e^{2\pi i t}, e^{2\pi i \alpha t})$$

for some irrational real number α. Then φ is an immersion of $\mathbb{R}$ in T^2 with dense image.

That an injective homomorphism is an immersion follows from properties of the exponential map, which we shall also introduce by means of examples below. The delicacy of the notion of a subgroup in this context is shown by the following proposition, which holds for $\mathcal{C}^0$- or topological groups.

Proposition 8.1.7. 1) *If the subgroup H of G contains a non-empty open subset then H is open, and* 2) *if H is open then H is closed.*

Proof. Suppose that $\emptyset \neq U \subseteq H$ is open. If $u \in U$, $u^{-1}U$ is an open neighbourhood of the identity e, and for each $h \in H$, h belongs to the open set $hu^{-1}U \subseteq H$ by the closure of multiplication. It follows that $H = \bigcup_{h \in H} hu^{-1}U$ and is open. For (2) note that $xH \cap H = \emptyset$ whenever $x \notin H$, and so $G \setminus H = \bigcup_{x \notin H} xH$, a union of open sets. As the complement of this open set, H itself is closed. ■

The closed subgroups of G are thus an important subclass among all the subgroups of G. However Example 8.1.6 illustrated the second possibility.

If we require that the image of the group H in G be a submanifold we have

Proposition 8.1.8. *Let H be a subgroup of G as a group, and a submanifold of G as a manifold. Then H is closed in G and is itself a Lie group with respect to the substructures.*

Proof. The smoothness of the operations restricted to H follows from its being a submanifold. In order to show that H is a closed subset of G use the definition of a submanifold to find a neighbourhood of e in G such that $U \cap H$ is closed relative to U. Choose a smaller neighbourhood V of e such that $V^{-1}V \subseteq U$. Now let $x \in \overline{H}$ and choose a sequence of points $\{h_n \in H\}$ with $\lim_{n\to\infty} h_n = x$. This is possible since the topology on G is locally Euclidean and first countable. Then there exists some n_0 with $h_n \in xV$ for all $n \geqslant n_0$, so in this range $h_n^{-1}h_{n_0} \in V^{-1}x^{-1}xV \subseteq U$. Hence $h_n^{-1}h_{n_0} \in U_n \cap H$ and $\lim_{n\to\infty}(h_n^{-1}h_{n_0}) = x^{-1}h_{n_0} \in V \subseteq U$. It follows that $x^{-1}h_{n_0} \in U \cap H$, since this is closed relative to U, so that $x \in H$. ■

Conversely it can be shown that a closed subgroup of a Lie group is an embedded submanifold, and hence a Lie subgroup. We start with a special case.

Theorem 8.1.9. *If $\varphi : G_1 \to G_2$ is a homomorphism of Lie groups, then φ has constant rank on G_1; thus $\ker(\varphi)$ is a (closed) embedded submanifold and so is a Lie subgroup with $\dim(\ker(\varphi)) = \dim(G_1) - \operatorname{rank}(\varphi)$.*

Proof. Let $a \in G_1$, $b = \varphi(a)$ and let e_1, e_2 be the identities of G_1, G_2, respectively. Then for all $x \in G_1$

$$\varphi(x) = \varphi(aa^{-1}x) = \varphi(a)\,\varphi(a^{-1}x) = (L_b \circ \varphi \circ L_{a^{-1}})(x).$$

Thus, the derivative $\varphi_*(a)$ of φ at a is

$$\varphi_*(a) = (L_b)_*(e_2) \circ \varphi_*(e_1) \circ (L_{a^{-1}})(a),$$

so that, since L_b and $L_{a^{-1}}$ are both diffeomorphisms, $\operatorname{rank}_a(\varphi) = \operatorname{rank}_{e_1}(\varphi)$. Thus φ has constant rank. The proof that $\ker(\varphi) = \varphi^{-1}(e_1)$ is a closed embedded submanifold follows from our earlier discussion of regular values. ■

8.2. Lie Algebras.

We used the left-invariant vector fields on a Lie group to show that the tangent bundle of a Lie group is trivial. In fact left-invariant vector fields play an even more fundamental rôle in the study and classification of Lie groups, which we shall now survey.

Let $\Gamma(\tau(M))$ denote the family of smooth sections of $\tau(M)$, the vector fields on M. Recall that if $f \in \mathcal{C}^\infty(M)$ and X is a vector field then we can regard $Xf(x)$ as the directional derivative of the function f at the point x. If X and Y are two vector fields, then we have already defined their *Lie bracket* $[X, Y]$ by the formula

$$[X, Y]f = X(Yf) - Y(Xf)$$

for all $f \in \mathcal{C}^\infty(M)$. When discussing the bracket operation on vector fields in Section 2.6 we obtained an expression for it in terms of local coordinates, viz.

$$[X, Y] = \sum_{i,j} \left(a_j \frac{\partial b_i}{\partial x_j} - b_j \frac{\partial a_i}{\partial x_j} \right) \frac{\partial}{\partial x_i}.$$

We have already noted in Section 2.5 that the properties there derived for vector fields and the Lie bracket make the space of smooth sections of the tangent bundle $\tau(M)$ into a real Lie algebra. For a general manifold this is an unwieldy infinite dimensional algebra, but for Lie groups it has a characteristic subalgebra of the same dimension as the group.

Definition 8.2.1. Let $\varphi : M^m \to N^n$ be smooth. The vector fields X on M and Y on N are said to be *φ-related* if $\varphi_* \circ X = Y \circ \varphi$.

Lemma 8.2.2. *If the pairs of vector fields $\{X_i, Y_i \mid i = 1, 2\}$ are φ-related, then $[X_1, X_2]$ and $[Y_1, Y_2]$ are also φ-related.*

Proof. For each $f \in \mathcal{C}^\infty(N)$ we need to show that

$$[X_1, X_2]_x(f \circ \varphi) = [Y_1, Y_2]_{\varphi(x)}(f),$$

which is done by unravelling the definitions. ■

Corollary 8.2.3. *If X and Y are left-invariant vector fields for the Lie group G, then $[X, Y]$ is also left-invariant.*

Proof. Apply Lemma 8.2.2 with $\varphi = L_x$. ■

This corollary shows that we may define the Lie Algebra of G to be the subalgebra of smooth sections of the tangent bundle formed by all left-invariant vector fields. If we denote it by $\mathfrak{G}$ we have already shown that $\mathfrak{G}$ is isomorphic as a real vector space to $\tau_e(G)$, the tangent space to the group at the identity. In this way $\tau_e(G)$ acquires a Lie algebra structure, and clearly $\dim(\mathfrak{G}) = \dim(\tau_e(G)) = \dim(G)$.

Examples 8.2.4. 1) Let $G = (\mathbb{R}^n, +)$. Then $\mathfrak{G}$ is the abelian Lie algebra $\mathbb{R}^n$ with all brackets equal to zero.

2) Let $G = GL(n, \mathbb{R})$. Then $\tau_e(G) = M_n(\mathbb{R}) \cong \mathbb{R}^{n^2}$, which describes $\mathfrak{G}$ as a vector space. To each $X \in \mathfrak{G}$ associate the $n \times n$ matrix $A = (a_{ij})$ of components of $X(e)$, so that $X(e) = \sum\limits_{ij} a_{ij} \left(\frac{\partial}{\partial x_{ij}}\Big|_e \right)$, and write $A = \mu(X)$. Then, by an explicit inspection of components, one can show that $\mu([X, Y]) = \mu(X)\,\mu(Y) - \mu(Y)\,\mu(X)$, giving the Lie algebra structure on $\mathfrak{G} = \mathfrak{G}\ell(n, \mathbb{R})$.

Let $\varphi : G \to H$ be a homomorphism of Lie Groups and denote their Lie algebras by $\mathfrak{G}$ and $\mathfrak{H}$ respectively. Define $\varphi_* : \mathfrak{G} \to \mathfrak{H}$ as follows: for each $X \in \mathfrak{G}$, $X(e) \in \tau_e(G)$ and the derivative $(\varphi_e)_*$ assigns $(\varphi_e)_*(X(e)) \in \tau_e(H)$ to the vector $X(e)$. Let $\varphi_*(X)$ be the unique element of $\mathfrak{H}$ taking the value $(\varphi_e)_*(X(e))$ at the identity of H; one can easily show that $\varphi_*(X)$ is the unique element of $\mathfrak{H}$ with $(\varphi_*(X)) \circ \varphi = \varphi_* \circ X$.

Proposition 8.2.5. 1) *If $\varphi : G \to H$ is a homomorphism of Lie groups, then $\varphi_* : \mathfrak{G} \to \mathfrak{H}$ is a homomorphism of Lie algebras.*

2) *The correspondence $G \to \mathfrak{G}$, $\varphi \mapsto \varphi_*$ is a covariant functor from the category of Lie groups to the category of Lie algebras.*

Proof. Granted Lemma 8.2.2 above, this is a manipulative exercise. ■

8.3. The Exponential Map.

Collecting together various facts we now have enough theory to show that the subgroups $SL(n, \mathbb{R})$ and $SO(n)$ of $GL(n, \mathbb{R})$ are Lie Groups, as are $U(n)$ and $SU(n)$ in $GL(n, \mathbb{C})$. In this section we give a heuristic discussion of their Lie algebras and also illustrate by means of two theorems how the Lie algebra of a group influences the group itself. These results hinge on the theorem that there is a $\mathcal{C}^\infty$-map $\exp : \mathfrak{G} \to G$, which in the case of the general linear group reduces to the classical exponential map on matrices:

$$\exp(A) = \lim_{k\to\infty} \sum_{i=0}^{k} \frac{1}{i!} A^i,$$

for an $n \times n$ matrix A. This limit exists for all A since the sequence on the right is easily seen to be a Cauchy sequence with respect to the Euclidean metric on $\mathbb{R}^{n^2}$.

A homomorphism $\varphi : \mathbb{R} \to GL(n, \mathbb{R})$ is called a *one-parameter* subgroup. It is necessarily of the form $\varphi(t) = \exp(tA)$, with $A = \varphi_*(0)$, since direct calculation shows that $\varphi_*(t) = \varphi_*(0)\varphi(t)$, and the unique solution of this equation with the initial condition $\varphi(0) = I$, the identity matrix, is $\varphi(t) = \exp(tA)$. In addition the map $\exp : M(n, \mathbb{R}) \to GL(n, \mathbb{R})$ is bijective near zero, its inverse being given by $g \mapsto \log g$, for $\|g - I\| < 1$. Here, as over the real or complex numbers

$$\log(I - A) = -\lim_{k\to\infty} \sum_{i=1}^{k} \frac{A^i}{i}.$$

We note that the complex exponential map has the following properties.

(*i*) $B \exp(A) B^{-1} = \exp(BAB^{-1})$.

(*ii*) $\det(\exp(A)) = e^{\mathrm{trace}(A)}$.

(*iii*) If A, B are such that $[A, B] = AB - BA = 0$, then $\exp(A + B) = \exp(A)\exp(B)$.

To see (ii), reduce A to a diagonal, or at worst a triangular, matrix and use (i). It follows from this that $\exp(A)$ is always non-singular, that is, that the exponential map takes values in $GL(n, \mathbb{C})$. From (iii) we see that, although in general exp is not a homomorphism from $(\mathfrak{G}, +)$ to $(G, \times)$, it does have this property on abelian subalgebras, in particular on the subalgebra $\{tA \mid t \in \mathbb{R}\}$ generated by A whose image is a 1-parameter subgroup. For example, $(\exp(A))^{-1} = \exp(-A)$. These properties, together with the trivial observations that $\exp(\bar{A}) = \overline{\exp(A)}$ and $\exp(A^T) = (\exp(A))^T$, where $\bar{A}$ is the complex conjugate and A^T the transpose of A, give us the following correspondences between subgroups of $G = GL(n, \mathbb{C})$ and subalgebras of $\mathfrak{G} = \mathfrak{G}l(n, \mathbb{C})$:

$$
\begin{aligned}
SL(n, \mathbb{C}) &\longleftrightarrow \{\text{ matrices of trace } 0\} \\
U(n) &\longleftrightarrow \{\text{ skew-Hermitian matrices}\} \\
O(n, \mathbb{C}) &\longleftrightarrow \{\text{ skew-symmetric matrices}\} \\
SU(n) &\longleftrightarrow \{\text{ skew-hermitian matrices of trace } 0\} \\
SL(n, \mathbb{R}) &\longleftrightarrow \{\text{ real } n \times n \text{ matrices of trace zero}\} \\
SO(n) &\longleftrightarrow \{\text{ real skew-symmetric matrices}\}.
\end{aligned}
$$

The reader should check that the sets of matrices on the right hand side are all closed with respect to the Lie bracket. The examples illustrate the general principle that, given a subgroup H of $GL(n, \cdot)$, we use the exponential map to translate the defining equations back to $\mathfrak{G}l(n, \cdot)$ to describe $\mathfrak{H}$.

Turning to an arbitrary Lie group G we have the following generalisation of the situation for matrix groups.

Theorem 8.3.1. *There is a* $1-1$ *correspondence between tangent vectors at the identity* $\tau_e(G)$ *and the* 1*-parameter subgroups* $\varphi : \mathbb{R} \to G$.

Proof. We argue as in the special cases of $GL(n, \mathbb{R})$ or $GL(n, \mathbb{C})$. A homomorphism φ gives us a tangent vector $\varphi_*(0) \in \tau_e(G)$.

Conversely, given $A \in \tau_e(G)$, we may integrate the tangent vector field X_A on G defined by $X_A(g) = (R_g)_*(A)$, where R_g is the right multiplication by g, subject to the initial condition $\varphi(0) = e$. At first glance it may appear that φ is only defined in some small open interval $(-\epsilon, \epsilon)$, but even if G is not compact the solution curve can be extended by means of the trick

$$\varphi(t) = \varphi(t/n)^n$$

for sufficiently large n. It is easy to check that this definition is independent of n, provided this is large, and $\varphi(t+u) = \varphi(t)\varphi(u)$ by the uniqueness of the solution to the ordinary differential equation. ■

The theorem implies that near $0 \in \tau_e(G)$ the tangent space is locally diffeomorphic to a neighbourhood of $e \in G$. If $\varphi = \varphi^A$ is the 1-parameter subgroup just defined, then we set

$$\exp(A) = \varphi^A(1) \in G.$$

To see that this map is a local diffeomorphism we note first that both the 1-parameter subgroups $s \mapsto \varphi^{tA}(s)$ and $s \mapsto \varphi^{A(ts)}$ are associated with $tA \in \mathfrak{G}$, and hence they are equal. Thus

$$\exp(tA) = \varphi^{tA}(1) = \varphi^A(t),$$

so that

$$\left.\frac{\partial}{\partial t}\right|_0 \exp(tA) = A.$$

But exp can be regarded as the restriction of the smooth flow $(t, g, A) \mapsto (g\varphi^A(t), A)$, obtained by integrating the vector field $(g, A) \mapsto (Ag, 0)$, to the subset $\{1\} \times \{e\} \times \mathfrak{G} \subseteq \mathbb{R} \times G \times \mathfrak{G}$. This map is smooth, its derivative at 0 is the identity, and so by the Inverse Function Theorem, it is locally invertible.

We are now in a position to present, at least in outline, a proof of the converse to Proposition 8.1.8, our first example of a property of Lie groups whose proof uses the Lie algebra and exponential map.

Theorem 8.3.2. *If H is a closed subgroup of the Lie group G, then H is a submanifold of G.*

Sketch proof. It suffices to find a submanifold structure near the identity element e of G. Give the Lie algebra $\mathfrak{G}$ a Euclidean metric, and choose a local inverse ('log') for the exponential map on the neighbourhood U of the identity e. Write $H' = \log(H \cap U)$. Using the assumption that H is closed we can find a *subspace* $W \subseteq \mathfrak{G}$ such that $\exp(W)$ is a neighbourhood of e in H. To be quite specific W is the set

$$\left\{ sX \,\middle|\, s \in \mathbb{R},\ X = \lim_{n\to\infty}\left(\frac{h_n}{\| h_n \|}\right) \text{ where } h_n \in H' \text{ is s.t. } \lim_n h_n = 0 \right\}.$$

We must show that exp is well-defined near $0 \in \mathfrak{G}$ as a map $W \to H$. Assume that the domain overlaps $W^\perp$ non-trivially and use a sequence (X_n, Y_n) in $\mathfrak{G}$

tending to 0 with $Y_n \neq 0$. We obtain a unit vector $Y \in W^{\perp}$ with $\exp(Y)$ belonging to the closed subgroup H, a contradiction. ■

Having considered closed subgroups H let us consider the quotient space G/H of left cosets with the quotient topology. Start by recalling some basic facts about transformation groups - the reader has already met examples in Chapter 3 and in Exercise 1.5 at the end of Chapter 1. A G-action on a space X may be thought of as a representation of G in the homeomorphism group of X, or in the diffeomorphism group if we restrict attention to smooth actions.

Definitions 8.3.3. For a group G acting on a set X, the *isotropy subgroup* at $x \in X$ is $G_x = \{g \in G \,|\, gx = x\}$. The *orbit* of x is $Gx = \{gx \mid g \in G\}$. The action is said to be *transitive* if it has exactly one orbit. When X is a topological space, the *orbit space* is the quotient space: the set of orbits topologised with the quotient topology.

We note that, with each orbit, we can associate a *conjugacy class* of isotropy subgroups, since the isotropy subgroup at $g(x)$ is gG_xg^{-1}, and that, although our notation is for left actions, we can equally well speak of right actions of G on X. A left action gives rise to a right action via the rule $xg = g^{-1}x$, and conversely.

The quotient or homogeneous space G/H, comprising cosets of the form xH, can be thought of as the space of orbits for the *right* action of H on G given by $(g, h) \mapsto gh$. The group G itself acts on G/H according to the rule, the *left* action,

$$G \times G/H \longrightarrow G/H; \quad (g, xH) \mapsto gxH.$$

We wish to show that, subject to the choices we have made, and reverting to the language of Chapter 3, G has the structure of a principal H-bundle over G/H. Note, before we start the main proof, that H operates on the total space G on the right, but that in changing charts we multiply by $g_{ij}(x) \in H$ on the left. These two actions commute.

Theorem 8.3.4. *Let G be a Lie group and H a closed subgroup of G. Then G admits the structure of an H-principal bundle with projection $p : G \to G/H$; $g \mapsto gH$, p is a submersion of rank equal to the dimension of G/H, and G/H inherits the structure of a differential manifold.*

Proof. (a) The quotient topology on G/H is Hausdorff. Let $xH \neq yH$, and let K be a compact neighbourhood of the point in x disjoint from the closed coset yH. Then KH is a closed neighbourhood of xH disjoint from yH, whose image $p(KH)$ suffices to separate the points in the orbit space.

(b) The local product structure of G depends on splitting $\mathfrak{G}$ as $V \oplus \mathfrak{H}$ (compare Theorem 8.3.2) and defining the *slice* D_E normal to H at e to be the image of the open E-disc in V about 0 under the exponential map. Then if ε is sufficiently small the natural map $D_\varepsilon \times H \to G$; $(g, h) \mapsto gh$ is an open embedding. The fact that the derivative is the identity on both factors provides us with an immersion near (e, e), which is injective by further reduction, if needed, in the value of ε.

(c) Cover G/H with the subsets $U_g/H = (gD_\varepsilon \cdot H)/H$, so that translates of the original slice provide the charts for an atlas on G/H. Globally the group G can now be expressed as the union of the local products $(U_g/H) \times H$, and the definition of U_g in terms of D_ε and the group action provides the H-principal structure. ■

As an important special case we note that, if H is a closed normal subgroup of G, then G/H is again a Lie group.

8.4. Maximal Tori and Cohomology.

As a final stop on this quick tour of Lie groups, we consider their de Rham cohomology. The most general result is due to Heinz Hopf and states that if G is a compact and connected Lie group, then $H^*(G)$ is isomorphic to the cohomology ring of a product of odd dimensional spheres. In terms of differential forms Hopf's Theorem states that there exists a family $\omega_1, \omega_2, \cdots, \omega_\ell$ of forms with degrees $m_1, m_2, \cdots, m_\ell$ having the following properties:

(i) For each k those exterior products $\omega_{i_1} \wedge \omega_{i_2} \wedge \cdots \wedge \omega_{i_r}$ for which $i_1 < i_2 < \cdots < i_r$ and $m_{i_1} + m_{i_2} + \cdots + m_{i_r} = k$ form an $\mathbb{R}$-linear basis for the invariant forms of degree k,

(ii) all the m_i are odd, and

(iii) the dimension of G is $m_1 + m_2 + \cdots + m_\ell$.

Note that by Theorem 6.5.1, if $m \geqslant 1$, $H^k(\mathbf{S}^m) = 0$ unless $k = 0$ or m, when the groups are isomorphic to $\mathbb{R}$. From Section 6.6 we also know that the de Rham cohomology of the m-dimensional torus T^m is generated as a ring by 1-dimensional classes, one associated with each factor $\mathbf{S}^1$. This proves Hopf's Theorem under the additional assumption that G is abelian, since tori are the only connected, compact, abelian Lie groups. This follows using the exponential map.

Further example: in Exercise 1.6.4 the reader was asked to prove that $\mathbb{R}\mathrm{P}^3$ is diffeomorphic to the group $SO(3)$. An easier variant of this exercise shows that there is an isomorphism of Lie groups $SU(2) \cong Sp(1) \cong \mathbf{S}^3$, where the 3-sphere is identified with the space of quaternions of unit length. Since the natural map $\mathbf{S}^3 \to \mathbb{R}\mathrm{P}^3$ is a double cover, and at worst can introduce 2-torsion into cohomology, we may conclude that

$$H^*(SO(3)) = H^*(SU(2))$$

is generated as above by a single class of forms in dimension three. Again this is in line with Hopf's Theorem. Using more than the methods described in this book it is possible to show that the result for abelian groups implies the result for more general G.

Definition 8.4.1. The subgroup $T \subset G$ is said to be a *maximal torus* if it is a torus, that is, $T \cong \mathbf{S}^1 \times \cdots \times \mathbf{S}^1$, and there is no other torus $T' \neq T$ with $T \subset T' \subset G$.

If G is compact and connected then maximal tori exist. Their importance is shown by

Theorem 8.4.2. *Any two maximal tori in G are conjugate, and every element of G is contained in a maximal torus.* ■

Closely associated to a maximal torus is the Weyl group $W(G)$, defined as the (finite) quotient group $N(T)/T$. Here $N(T)$ is the *normaliser* of T in G, that is

$$N(T) = \{g \in G \,|\, gTg^{-1} = T\}.$$

Familiar examples are provided by:

1) In the case of $SU(n)$ the maximal torus consists of diagonal matrices $\mathrm{diag}\{\lambda_1, \cdots, \lambda_n\}$ such that $\lambda_1\lambda_2 \cdots \lambda_n = 1$. The Weyl group $W(SU(n))$ equals the symmetric group $S(n)$, identified with the group of permutation matrices.

2) For the orthogonal groups $SO(2n)$ and $SO(2n+1)$ we have the inclusions $T(n) = SO(2) \times \cdots \times SO(2) \subset SO(2n) \subset SO(2n+1)$, $T(n)$ providing a maximal torus in both cases. $W(SO(2n+1))$ is obtained by allowing $S(n)$ to act on $(\mathbb{Z}/2)^n$ by permuting the factors, also allowing ± 1 as entries in the permutation matrices. If we call this group $G(n)$ then $SG(n)$, corresponding to the even permutations, equals $W(SO(n))$. As in Chapters 5 and 6, the smooth inclusion homomorphism $T \subset G$ induces the restriction homomorphism between de Rham groups

$$H^*(G) \longrightarrow H^*(T).$$

Furthermore the image is contained in the subset of elements invariant under the action of the normaliser $N(T)$. This turns out to be enough to describe $H^*(G)$. In the special cases of the three orthogonal and special unitary groups the dimensions of the spheres $\mathbf{S}^{m_i}$ arising in Hopf's Theorem are $m_i = 2d_i + 1$ where for

$$\begin{aligned} SU(n): &\quad (d_1, \cdots, d_{n-1}) = (1, 2, \cdots, n-1); \\ SO(2n): &\quad (d_1, \cdots, d_n) = (1, 3, \cdots, 2n-3, n-1); \\ SO(2n+1): &\quad (d_1, \cdots, d_n) = (1, 3, \cdots, 2n-1). \end{aligned}$$

The integers d_i are known as the *exponents* of the groups concerned. As a final check we note that for $SU(2)$ we have $m_1 = 3$, for $SO(2) = \mathbf{S}^1$ we have $m_1 = 1$, and for $SO(3)$ we have $m_1 = 3$, confirming our previous calculations.

8.5. Exercises for Chapter 8.

8.1. Establish the claims in Examples 8.2.4
  (*i*) that the Lie algebra of $(\mathbb{R}^n, +)$ is the abelian Lie algebra $\mathbb{R}^n$;
  (*ii*) that $\mu([X,Y]) = \mu(X)\mu(Y) - \mu(Y)\mu(X)$ in the notation given there.

8.2. Show that, if K is a compact neighbourhood of x in a Lie group and K is disjoint from the coset yH of the closed subgroup H of G, then KH is a closed neighbourhood of xH disjoint from yH.

8.3. Use the identification of $SU(2)$ with the unit quaternions to define an action of $SU(2) \times SU(2)$ on the quaternions by $q \mapsto aqb^{-1}$. Deduce that $SU(2) \times SU(2)$ is the double cover, Spin(4), of $SO(4)$.

CHAPTER 9

A RAPID COURSE IN DIFFERENTIAL ANALYSIS

Prerequisites for a Course on Differential Manifolds

This chapter is intended only as a reminder, not as a first course on the topics mentioned. Indeed a thorough understanding of these topics, while very desirable, is not what is strictly required. What is necessary is that the reader should feel comfortable with the concepts mentioned in the main text, and have some idea how they behave, and not that the authors are talking about some mysterious object which has little meaning. Should the latter feeling surface at any time, a glance through the relevant section of this chapter may help to cure it.

Although the previous chapters deal exclusively with finite dimensional Euclidean spaces, we include here a number of comments, and even a few proofs, on potentially infinite dimensional normed spaces and hence, by implication for the rest of the book, on infinite dimensional manifolds. If the reader feels overwhelmed by this it is quite adequate just to read $\mathbb{R}^m$ or $\mathbb{R}^n$ for $\mathbb{E}$ or $\mathbb{F}$.

The majority of the prerequisites are from elementary analysis. However there are three further topics that, like the inverse function theorem, tend to fall into the gaps between many university courses. Although they are algebraic rather than analytic, it seems sensible to include them at the end of this chapter.

9.1. Metric Spaces.

In order to do analysis, in other words in order to perform processes involving the taking of limits, we need some means of telling how close to that limit we are. This leads us to the concept of a metric or measure of distance. In a vector space we require this metric to tie in with the linear structure in a manner which produces what is called a norm.

Definition 9.1.1. 1) A *metric* on a set S is a mapping $d : S \times S \to \mathbb{R}^+$ from the cartesian product of S with itself to the non-negative reals, satisfying the following axioms:

(*i*) $\forall\, a,\, b \in S,\ d(a,\, b) = d(b,\, a)$;

(*ii*) $\forall\, a,\, b \in S,\ d(a,\, b) = 0 \Longleftrightarrow a = b$;

(*iii*) $\forall\, a,\, b,\, c \in S,\ d(a,\, c) \leqslant d(a,\, b) + d(b,\, c)$.

2) A *metric space* is a set S together with a metric d upon it.

In any metric space, we denote by $B(x,\, \delta)$ the *open ball centre x and radius δ*, that is, the set $\{y \,|\, d(x, y) < \delta\}$, and by $\overline{B(x,\, \delta)}$ the corresponding *closed ball*, $\{y \,|\, d(x, y) \leqslant \delta\}$.

We now define a mapping between metric spaces to be continuous, uniformly continuous etc., using ϵ–δ-technology, just as we did in $\mathbb{R}$, but now using $d(x, y)$ in place of $|x - y|$. Except for this the words and symbols are the same, but they refer to a much wider class of objects. Moreover the axioms we have required for a metric ensure that most of our proofs are still valid in the new context. Similarly, in a metric space X we define convergence and uniform convergence of sequences, Cauchy sequences, etc., just as for $\mathbb{R}^1$ but again with $d(x, y)$ in place of $|x - y|$. Then a metric space is *sequentially compact* if every sequence has a convergent subsequence. It is *complete* if every Cauchy sequence converges and it is *bounded* if there is a d such that $d(x, y) < d$ for all x, y in X; the least such d being the *diameter of* X, $\mathrm{diam}(X)$.

Note that convergence implicitly requires that the sequence converge to a point of the space itself not, as may be the case, to a point of some larger space.

The local models of the metric spaces with which differential geometry is concerned carry a linear algebraic structure. We shall assume without further comment the basic properties of vector spaces over $\mathbb{R}$ or $\mathbb{C}$, such as the Exchange Lemma for bases, the representation of linear maps by matrices and, most importantly, the properties of dual bases and dual linear transformations. The reader should also be familar with the elementary theory of bilinear forms, especially those satisfying a symmetric or skew-symmetric condition. A good reference is Jänich (1994).

Definition 9.1.2. A *norm* on a real or complex vector space V is a mapping $\|\cdot\| : V \to \mathbb{R}^+$ from V to the non-negative reals, satisfying the following axioms:

(*i*) $\forall\, v \in V,\ \forall\, \lambda \in \mathbb{R},\ \|\lambda v\| = |\lambda|\ \|v\|$;

(*ii*) $\forall v \in V,\ \|v\| = 0 \iff v = 0$;
(*iii*) $\forall v,\ w \in V,\ \|v + w\| \leqslant \|v\| + \|w\|$.

We shall generally write $\mathbb{V}$ for the vector space V equipped with a norm. If V has dimension m we often refer to a ball in V as an *m-ball* and denote the ball of unit radius centred on the origin, $B(0,\ 1)$, simply by B^m.

Examples 9.1.3. The most basic example of a metric space is $(\mathbb{R},\ d)$, the real numbers with the metric d defined by $d(a,\ b) = |a - b|$. More generally any subset of $\mathbb{R}$ with the restriction of this metric is a metric space. Generalising in another direction, if $\|\cdot\|$ is a norm on the vector space $\mathbb{V}$ then $d(a,\ b) = \|a - b\|$ is a metric on the underlying set V. In particular, from the Euclidean norm defined by $\|x\|^2 = \sum_{i=1}^{k} x_i^{\,2}$ on $\mathbb{R}^k$, we get, by this means, the Euclidean metric on $\mathbb{R}^k$.

Definition 9.1.4. A normed vector space which is complete with respect to the induced metric is called a *Banach space.*

Remarks 9.1.5. Much of the content of these notes may be adapted to Banach spaces which may be infinite dimensional and much current research does involve infinite dimensional manifolds. Do not succumb to the impression that one has to work in finite dimensions. Nevertheless the adaption to infinite dimensions is sometimes subtle and one's intuition certainly lies in very low finite dimensions, and in this book we only consider finite dimensional $\mathbb{V}$ where the important facts are:

(*i*) All norms are *equivalent* that is they induce the same topology, though not necessarily the same metric, which means that they determine the same continuous maps and convergent sequences etc. Most importantly for our present purposes, which maps are differentiable, and what their derivatives are, does not depend on the choice of norm.

(*ii*) All linear maps $f : \mathbb{V} \to \mathbb{W}$ are continuous and indeed, as we shall see, differentiable.

Thus in practice we shall lose no generality if

we think of our spaces as $\mathbb{R}^n$ with the Euclidean norm.

Note that spaces of linear maps, being themselves linear spaces may also be given norms. We often take the sup of the moduli of the entries of a matrix representation as the norm. For normed spaces $\mathbb{V}$ and $\mathbb{W}$, with $\mathbb{V}$ finite-dimensional, we may always define the *operator norm* on $\mathcal{L}(\mathbb{V}, \mathbb{W})$ by

$\|\alpha\| = \sup_{\|v\|=1} \|\alpha(v)\|$. This has the property that, for any $v \in \mathbb{V}$ and $\alpha \in \mathcal{L}(\mathbb{V}, \mathbb{W})$, $\|\alpha(v)\| \leqslant \|\alpha\| \, \|v\|$. For example, for a linear map between Euclidean spaces with Euclidean norms the operator norm is just the maximum of the norms of the images of the standard orthonormal basis vectors or, equivalently, the maximum of the norms of the columns of the matrix that (operating on the left) represents the map with respect to the standard bases.

As with vector spaces over $\mathbb{R}$ or $\mathbb{C}$, we also assume some familiarity with the basic definitions from the theory of topological spaces, having in the back of our minds the inclusions

$$\text{inner product spaces} \subseteq \text{normed spaces} \subseteq \text{metric spaces} \subseteq \text{topological spaces.}$$

We regard topologies on the same set X as being partially ordered by inclusion ('finer' versus 'coarser'), and use the separation properties for isolated points and closed subsets (Hausdorff, regular, normal, etc.), all of which are satisfied by metric spaces. The space $(X, \mathcal{T})$ is *compact* if every covering of X by open sets has a finite subcover, and *locally compact* if each point $x \in X$ has a neighbourhood for which this holds. $(X, \mathcal{T})$ is *second countable* if each open subset is the union of members from some preferred countable subfamily, and is *disconnected* if it is the union of two disjoint open, non-empty subsets. Quotient spaces are particularly important in geometric constructions, and the reader should certainly be familiar with these, see for example the first half of Armstrong (1997). This is also a good reference for the other material referred to. It is non-trivial, and very rewarding, to work out conditions under which each of the properties above is inherited by subspaces, products and quotient spaces. For example, not every subset of a compact space need be compact, however every closed subset will be. This result, together with the facts that the continuous image of a compact space is compact and that a compact subset of a Hausdorff space is closed, establish the following useful result.

Lemma 9.1.6. *A continuous injection f from a compact space to a Hausdorff space is a homeomorphism onto its image.*

Proof. Since the inverse mapping is certainly defined on the image, to show that it is continuous it suffices to show that f itself maps closed sets to closed sets. However if A is closed in a compact space then it is also compact and, hence, its continuous image $f(A)$ is compact. But $f(A)$ lies in a Hausdorff space and so is closed. ■

9.2. Contraction Mappings.

These are definable on metric spaces and play an important rôle in existence and uniqueness theorems. Thus to use them to solve a problem we need to put a metric on the set from which the solution is to be selected and recast the problem as a contraction of that space.

Definition 9.2.1. A *contraction mapping* of a metric space (X, d) into another (Y, e) is a mapping f such that there exists k in $[0, 1)$ such that for all x, y in X, $e(f(x), f(y)) < k\, d(x, y)$.

Usually we are only interested in contraction mappings of a space into itself. Note that $k < 1$ is essential, although the contraction condition may be weakened to $d(f(x), f(y)) \leqslant k\, d(x, y)$. In particular the identity is not a contraction. However if X is not compact then, contrary to naive intuition, a contraction mapping may be surjective; for example, $f\colon \mathbb{R}^1 \to \mathbb{R}^1; x \mapsto kx$ with $k < 1$. The main theorem is the following.

Theorem 9.2.2. (Contraction Mapping Theorem.) *A contraction mapping of a complete metric space into itself has a unique fixed point.*

Proof. The uniqueness only requires $k \leqslant 1$, for if $x = f(x)$ and $y = f(y)$ are distinct fixed points then $d(x, y) = d(f(x), f(y)) < k\, d(x, y) \leqslant d(x, y)$ which is impossible. Existence proceeeds by showing that the sequence $x_{n+1} = f(x_n)$, for any x_0, is a Cauchy sequence, which does require $k < 1$. Then, by completenes the sequence has a limit x and $f(x) = x$ by the continuity of f. Thus in some sense the repeated applications of f 'contract X onto the unique fixed point'. ■

The main application in analysis is to the solution of differential and integral equations and similar problems. Thus the fixed point we require will be a function and so X must be a function space, a space whose points are functions between two other spaces. Since we required the solution of differential equations at various points in this book, we give here a typical example. A map between open subsets of Euclidean spaces is $\mathcal{C}^1$, respectively smooth, if the partial derivatives of all its coordinate functions are $\mathcal{C}^1$, respectively smooth.

Theorem 9.2.3. *Let U be open in $\mathbb{R}^m$, $I_a = (-a, a) \subseteq \mathbb{R}$ and $f : U \times I_a \to \mathbb{R}^n$ be of class $\mathcal{C}^r$ for $r \geqslant 1$. Then for all $x_0 \in U$, there exists $b > 0$ and $V \subseteq U$*

with $x_0 \in V$ *together with a unique* C^r*-mapping* $F : V \times I_b \to U$ *such that*

$$\begin{aligned} &F(x,0) = x \text{ for all } x \in V \quad \text{and} \\ &\frac{\partial F}{\partial t}(x,t) = f(F(x,t),\, t) \text{ for all } (x,t) \in V \times I_b. \end{aligned} \tag{9.1}$$

Proof. Choose $\delta > 0$ such that the closed ball $\overline{B(x_0, 2\delta)}$ is contained in U and $b > 0$ such that $\bar{I}_b = [-b, b]$ is contained in I_a. We now consider the space $\mathcal{M}$ of all continuous mappings $F : \overline{B(x_0, \delta)} \times \bar{I}_b \to \overline{B(x_0, 2\delta)}$ such that $F(x, 0) = x$, where b is still to be chosen. We give the maps F in $\mathcal{M}$ the 'sup' norm, $\|F\| = \sup_{(x,t)} \|F(x,t)\|$, where (x, t) ranges over the domain $\overline{B(x_0, \delta)} \times \bar{I}_b$. Then $\mathcal{M}$ is complete under the associated metric $d(F, G) = \|F - G\|$ and we define the map ϕ of $\mathcal{M}$ into itself by

$$(\phi(F))\,(x,t) = x + \int_0^t f(F(x,\tau),\tau)\,d\tau,$$

To see that $\phi(F) \in \mathcal{M}$, note first that $(\phi(F))\,(x, 0) = x$ and that $\phi(F)$ is continuous. Then, for $(x,t) \in \overline{B(x_0, \delta)} \times \bar{I}_b$,

$$\begin{aligned} \|(\phi(F))\,(x,t) - x\| &= \left\| \int_0^t f(F(x,\tau),\tau)\,d\tau \right\| \\ &\leqslant \int_0^t \|f(F(x,\tau),\tau)\|\,d\tau \\ &\leqslant b \times \sup\{\|f(y,t)\| \mid (y,t) \in \overline{B(x_0, 2\delta)} \times \bar{I}_a\} = b\,M_1, \end{aligned}$$

say. So we require $b\,M_1 < \delta$, that is, $b < \delta/M_1$.

By the Mean Value Theorem and the continuity of the derivative of f there is also a bound M_2 such that for all x, y in $\overline{B(x_0, 2\delta)}$ and $t \in \bar{I}_b$

$$\|f(x,t) - f(y,t)\| \leqslant M_2\|x - y\|.$$

Then,

$$\begin{aligned} \|\phi(F) - \phi(G)\| &\leqslant \int_0^t \|f(F(x,\tau),\tau) - f(G(x,\tau),\tau)\|\,d\tau \\ &\leqslant b\,M_2 \sup_{(x,t)} \|F(x,\tau) - G(x,\tau)\| \\ &= b\,M_2\, d(F, G) \end{aligned}$$

and, for ϕ to be a contraction mapping, we require $b\,M_2 < 1$, which we achieve by taking $b < 1/M_2$ if that is not already the case.

Thus, there is a unique $F \in \mathcal{M}$ such that $\phi(F) = F$, that is,

$$F(x,t) = x + \int_0^t f(F(x,\tau),\tau)\,d\tau$$

from which it follows that, in addition to $F(x,0) = x$,

$$\frac{\partial F}{\partial t}(x,t) = f(F(x,t),t),$$

so solving uniquely the given differential equation.

It follows that the solution F is $\mathcal{C}^{r+1}$ in the t variable. However, differentiability with respect to the other variables, the 'initial condition' x, requires more analysis and we shall not give the details here but refer readers to Dieudonné (1969) and Lang (1995). ■

9.3. Differential Analysis.

A function $f\colon \mathbb{R}^1 \to \mathbb{R}^1$ is linear if and only if $f(t) = \lambda t$ for some $\lambda \in \mathbb{R}$. Saying that f is differentiable at $x \in \mathbb{R}^1$ means that there exists $\lambda = f'(x)$ such that

$$f(x+t) - f(x) = t\lambda + o(t),$$

where $o(t)$ represents a function of the form $t\epsilon(t)$ with $\epsilon(t) \to 0$ as $t \to 0$. That is, near x the map $t \to f(x+t) - f(x)$ is approximated by the linear map $t \to tf'(x)$. We can use the same definition of differentiability for maps $f\colon \mathbb{R}^n \to \mathbb{R}^m$ provided we have a metric to measure the approximation which is properly tied in with the linear structure, that is, it is a norm.

On the other hand, we may define $f\colon \mathbb{R}^n \to \mathbb{R}$ to be *differentiable* if and only if all its partial derivatives exist and are continuous; and then $f\colon \mathbb{R} \to \mathbb{R}^n$ to be *differentiable* if and only if all its coordinate functions are differentiable.

Thus a differentiable map $f\colon \mathbb{R}^n \to \mathbb{R}^m$ acquires an $m \times n$ matrix $J(f)$ of partial derivatives of the coordinate functions, where $J(f)_{ij} = \partial f_i/\partial x_j$. In fact:

(*i*) these two approaches are closely related. For example, $J(f)$ is the matrix, with respect to the standard basis, of the linear map $f'(x)$ approximating f at x:

(*ii*) differentiable maps $f\colon \mathbb{R}^n \to \mathbb{R}^m$ share many simple properties with those for which $n = m = 1$:

(*iii*) two major theorems, the Mean Value Theorem and the Inverse Function Theorem, generalise to this wider context.

Definition 9.3.1. Let $f: U \to \mathbb{F}$, where U is open in $\mathbb{E}$ and $\mathbb{E}, \mathbb{F}$ are normed vector spaces. Then f is *differentiable at x in U* if there is a linear map $\phi: \mathbb{E} \to \mathbb{F}$ such that $f(x+h) - f(x) = \phi(h) + o(h)$, where $o(h)$ denotes a function, from a deleted neighbourhood of the origin in $\mathbb{E}$ into $\mathbb{F}$, such that $o(h)/\ \|h\| \to 0$ as $h \to 0$. Here ϕ is called the *derivative of f at x*, denoted $f'(x)$ or $Df(x)$. We shall use the latter to remind us that it is a linear map and not an element of $\mathbb{E}$ or $\mathbb{F}$. We say that f is *differentiable* or *differentiable on U* if it is differentiable at x for all $x \in U$.

If f is differentiable, then Df defines a map from U to $\mathcal{L}(\mathbb{E}, \mathbb{F})$. However $\mathcal{L}(\mathbb{E}, \mathbb{F})$, as already noted, may be given a norm and, by Remark 9.1.5(i) above, it does not matter which norm. Then, if Df is continuous, we say that f is $\mathcal{C}^1$, or of *class* $\mathcal{C}^1$, on U. If Df is differentiable, we say the f itself is *twice differentiable* and, if Df is $\mathcal{C}^1$, we say that f is $\mathcal{C}^2$. Inductively we define f to be $\mathcal{C}^r$ if Df is $\mathcal{C}^{r-1}$. If f is $\mathcal{C}^r$ for all r, then f is said to be $\mathcal{C}^\infty$ or *smooth.* We write $\mathcal{C}^r(U, \mathbb{F})$ for the set (in fact a vector space, see below) of $\mathcal{C}^r$-maps from U into $\mathbb{F}$. When $\mathbb{F} = \mathbb{R}$ is 1-dimensional we usually just write $\mathcal{C}^r(U)$.

Remark 9.3.2. Note that the second derivative $D(Df)(x) = D^2 f(x)$ is in $\mathcal{L}(\mathbb{E}, \mathcal{L}(\mathbb{E}, \mathbb{F}))$ which is naturally isomorphic with $\mathrm{Bil}(\mathbb{E} \times \mathbb{E}, \mathbb{F})$, the space of bilinear maps from $\mathbb{E} \times \mathbb{E}$ to $\mathbb{F}$. It is in this latter space that we usually think of the second derivative as lying. Similarly the nth derivative $D^n f(x)$ is usually regarded as being in the space of n-linear maps $\mathrm{Multilin}(\mathbb{E} \times \cdots \times \mathbb{E}, \mathbb{F})$.

Examples 9.3.3. 1) If $f: U \to \mathbb{F}$ is constant and $f(x) = p$ for all $x \in U$, then $Df(x) = 0$ for all x.

2) If $f: U \to \mathbb{F}$ is linear, then $Df(x) = f$ for all x. Note that, for the classical case $\mathbb{E} = \mathbb{F} = \mathbb{R}^1$, the linear map $f(x) = \lambda x$ has $f'(x) = \lambda$. When reinterpreted as a linear map $Df(x)$, this becomes $t \to \lambda t$, that is, f itself.

3) If $\beta: \mathbb{E} \times \mathbb{F} \to \mathbb{G}$ is bilinear, then $D\beta(e, f) \in \mathcal{L}(\mathbb{E} \oplus \mathbb{F}, \mathbb{G})$ is the linear map $(h, k) \mapsto \beta(e, k) + \beta(h, f)$, where $\mathbb{E} \oplus \mathbb{F}$ is the Cartesian product $\mathbb{E} \times \mathbb{F}$ with 'external direct sum' vector space structure.

4) For $f, g: U \to \mathbb{F}$ differentiable and λ, $\mu \in \mathbb{R}$ we may, as usual, define $\lambda f + \mu g$ pointwise by $(\lambda f + \mu g)(x) = \lambda f(x) + \mu g(x)$. Then $\lambda f + \mu g$ is differentiable with $D(\lambda f + \mu g)(x) = \lambda Df(x) + \mu Dg(x)$ where, on the right hand side, we have the usual (pointwise) linear combination of linear maps. Thus $\mathcal{C}^r(U, \mathbb{F})$ is a vector space.

5) For $f, g: U \to \mathbb{R}$ differentiable we may also define the product pointwise

by

$$fg: U \to \mathbb{R}; x \mapsto f(x)g(x).$$

Then fg is differentiable and

$$D(fg)(x) = f(x)\, Dg(x) + g(x)\, Df(x)$$

where again the right hand side is the usual linear combination. This, subject to checking a few identities, says that $\mathcal{C}^r(U) = \mathcal{C}^r(U, \mathbb{R})$ is an algebra. (See Section 9.6 below).

6) If $\mathbb{F} = \mathbb{F}_1 \times \mathbb{F}_2$ and $f_i: U \to \mathbb{F}_i$ are $\mathcal{C}^r$ then so too is $f: U \to \mathbb{F}$ defined by $f(x) = (f_1(x), f_2(x))$. Moreover $Df(x)$ is identified with $(Df_1(x), Df_2(x))$ under the natural isomorphism of $\mathcal{L}(\mathbb{E}, \mathbb{F})$ with $\mathcal{L}(\mathbb{E}, \mathbb{F}_1) \times \mathcal{L}(\mathbb{E}, \mathbb{F}_2)$.

If, for some open $U \subseteq \mathbb{E}$, $f: U \to \mathbb{F}$ is $\mathcal{C}^r$ and bijective onto $f(U)$, then f is called a *$\mathcal{C}^r$-diffeomorphism* if $f(U)$ is also open and the inverse $f^{-1} : f(U) \to U$ is also $\mathcal{C}^r$. More generally, f is called a *local $\mathcal{C}^r$-diffeomorphism at x* if there is an open neighbourhood U_x of x such that $f|_{U_x}$ is a $\mathcal{C}^r$-diffeomorphism. Note that this requires that $f(U_x)$ be open in the image space.

Theorem 9.3.4. (The Chain Rule.) *Let U, V be open in $\mathbb{E}, \mathbb{F}$ respectively, $f: U \to \mathbb{F}$ be differentiable at x in U with $f(x) \in V$ and $g: V \to \mathbb{G}$ be differentiable at $f(x)$. Then the composite $g \circ f$, defined on $U \cap f^{-1}(V)$, is differentiable at x with derivative*

$$D(g \circ f)(x) = Dg(f(x)) \circ Df(x),$$

where the composition on the right hand side is that of linear maps.

Proof. To simplify notation, write $f(x) = y$, $Df(x) = \alpha$, $Dg(y) = \beta$ and consider $x_1 \in U$. Then $y_1 = f(x_1) = y + \alpha(x - x_1) + \xi(x_1)\|x_1 - x\|$ where $\xi(x_1) \to 0$ as $x_1 \to x$. Similarly $g(y_1) = g(y) + \beta(y_1 - y) + \eta(y_1)\|y_1 - y\|$ where, without loss of generality, we may define $\eta(y) = 0$ and so have η continuous at y. Hence $g \circ f(x_1) - g \circ f(x) - \beta \circ \alpha(x_1 - x) = g(y_1) - g(y) - \beta(\alpha(x_1 - x)) = \beta(\xi(x_1)\|x_1 - x\|) + \eta(y_1)\|y_1 - y\| = \delta(x)\|x_1 - x\|$ where, for the last equality to hold, $\delta(x_1)$ must be $\beta(\xi(x_1)) + \eta(f(x_1))\{\alpha(x_1 - x)/\|x_1 - x\| + \xi(x_1)\}$. So $\delta(x_1)$ does indeed tend to zero as $x_1 \to x$ and the result is established. ■

Addendum 9.3.5. *If f, as above, is n-times differentiable at x and g is n-times differentiable at $f(x)$, then $g \circ f$ is n-times differentiable at x. Similarly $f \in \mathcal{C}^n(U, \mathbb{F})$, $f(U) \subseteq V$ and $g \in \mathcal{C}^n(V, \mathbb{G})$ implies $g \circ f \in \mathcal{C}^n(U, \mathbb{G})$.*

Proof. The proof is by induction, starting with the chain rule itself and writing the derivative $D(g \circ f): U \to \mathcal{L}(\mathbb{E}, \mathbb{G})$ as the composite

$$x \mapsto (Dg(f(x)),\ Df(x)) \mapsto D(g \circ f)(x),$$

where the second map is the bilinear, and hence $\mathcal{C}^\infty$, map

$$\mathcal{L}(\mathbb{F}, \mathbb{G}) \times \mathcal{L}(\mathbb{E}, \mathbb{F}) \longrightarrow \mathcal{L}(\mathbb{E}, \mathbb{G}); \quad (\alpha, \beta) \mapsto \alpha \circ \beta.$$

The first component of the first map is a composition of a $\mathcal{C}^n$-map and a $\mathcal{C}^{n-1}$ one, so is $\mathcal{C}^{n-1}$ by induction. The remaining details are straightforward. ■

Example 9.3.6. The norm on an inner product space $\mathbb{E}$ can be written as the composite of three maps, the linear map $\Delta : \mathbb{E} \to \mathbb{E} \times \mathbb{E}; x \mapsto (x,\, x)$ (usually called the *diagonal map*) followed by the bilinear inner product and then the square root. The first two of these are differentiable everywhere and the third everywhere except at zero so by, two applications of, the chain rule the norm is differentiable everywhere except at the origin.

9.3.1. Recovery of the Jacobian.

Let $\mathbb{E} = \mathbb{E}_1 \times \cdots \times \mathbb{E}_n$ and $\mathbb{F} = \mathbb{F}_1 \times \cdots \times \mathbb{F}_m$, direct products with the sup norm. Then we have the injections $\iota_j: \mathbb{E}_j \to \mathbb{E}; x_j \mapsto (0, \cdots, 0, x_j, 0, \cdots, 0)$ and projections $\pi_i: \mathbb{F} \to \mathbb{F}_i; (y_1, \cdots, y_m) \mapsto y_i$. Then, if $U \subseteq \mathbb{E}$ and $f: U \to \mathbb{F}$, the composite $f_i = \pi_i \circ f$ is the *ith coordinate function* of f, and $f(x) = (f_1(x), \cdots, f_m(x))$ as usual.

If $T: \mathbb{E} \to \mathbb{F}$ is linear it determines, and is determined by, the matrix of linear transformations $(T_{ij}) = (\pi_i \circ T \circ \iota_j)$. In the special case $\mathbb{E}_i = \mathbb{F}_j = \mathbb{R}$, when $\mathbb{E} = \mathbb{R}^n$ and $\mathbb{F} = \mathbb{R}^m$ we may replace the linear map $T_{ij}: \mathbb{R} \to \mathbb{R}$ by the multiplying scalar which determines it. Then we simply get the usual matrix representation for $T \in \mathcal{L}(\mathbb{R}^n, \mathbb{R}^m)$ with respect to the standard basis.

Then for U open in $\mathbb{E}$ and $f: U \to \mathbb{F}$ differentiable we define the *jth partial derivative of f at x* to be $Df(x) \circ \iota_j$ and usually denote it by $D_j f(x)$. For example when $\mathbb{F} = \mathbb{R}$ and $\mathbb{E} = \mathbb{R}^n$ we have the usual partial derivative $\partial_j f \equiv \partial f / \partial x_j = Df(x) \circ \iota_j \cdot 1$, since $f(x_1 \cdots, x_{j-1}, x_j + h, x_{j+1}, \cdots, x_n) - f(x) = Df(x) \cdot (\iota_j \cdot h) + o(|h|) = D_j f(x) \cdot h + o(|h|) = h \frac{\partial f}{\partial x_j}(x) + o(|h|)$, giving the more familiar definition.

The following derivation of the Jacobian matrix is a simple application of the chain rule and the fact that, since π_i is linear, $D\pi_i = \pi_i$ everywhere.

$$\begin{aligned} D_j f_i(x) &= Df_i(x) \circ \iota_j = D(\pi_i \circ f)(x) \circ \iota_j \\ &= D\pi_i(f(x)) \circ Df(x) \circ \iota_j \\ &= \pi_i \circ Df(x) \circ \iota_j = (Df(x))_{ij}, \end{aligned}$$

the (i,j)th element of the matrix which determines $Df(x)$. In general this is a linear transformation from $\mathbb{E}_j$ to $\mathbb{F}_i$ but, as mentioned above, in the particular case $\mathbb{E}_j = \mathbb{F}_i = \mathbb{R}$ we can replace it by the scalar which determines it and we then recover the Jacobian matrix $J(f(x))$ of partial derivatives of the coordinate functions, $\left(\frac{\partial f_i}{\partial x_j}\right)$ or, as it is commonly written $\frac{\partial(f_1,\cdots,f_m)}{\partial(x_1,\cdots,x_n)}$.

Theorem 9.3.7. (The Mean Value Theorem.) *Let U be open in the normed space $\mathbb{E}$ and $f \in \mathcal{C}^1(U,\mathbb{F})$. If $[a,b] = \{(1-t)a + tb \mid t \in [0,1]\} \subseteq U$ and, for all $x \in [a,b]$, $\| f'(x) \| \leqslant M$, then $\| f(b) - f(a) \| \leqslant M\|b-a\|$.*

Proof. For all $\epsilon > 0$ and for each x in $[a,b]$, since f is differentiable at x there is a δ_x such that $t \in \mathrm{B}(x,\delta_x) \cap U$ implies $\|f(t) - f(x) - f'(x)\cdot(t-x)\| \leqslant \epsilon\,\|t-x\|$. So $\| f(t) - f(x) \| \leqslant (M+\epsilon)\|t-x\|$ for all $t \in \mathrm{B}(x,\delta_x) \cap U$. Since $[a,b]$ is closed, bounded and hence compact it is covered by finitely many $\mathrm{B}(x,\delta_x)$, so we have a finite sum of inequalities giving $\| f(b) - f(a) \| \leqslant (M+\epsilon)\|b-a\|$, where we have used the fact that on the right hand side the norms are strictly additive since the vectors are collinear. Now ϵ is arbitrary so $\| f(b) - f(a) \| \leqslant M\|b-a\|$. ∎

Addendum 9.3.8. If we apply the Mean Value Theorem to $g(x) = f(x) - Df(a)\cdot x$ for $x \in [a,b]$, we get

$$\| f(b) - f(a) - Df(a)\cdot(b-a) \| \leqslant M\|b-a\|,$$

where $M = \sup\{ \| Df(x) - Df(a) \| \mid x \in [a,b]\}$.

Theorem 9.3.9. (Symmetry of the second derivative.) *Let U be open in $\mathbb{E}$ and $f \in \mathcal{C}^2(U,\mathbb{F})$. Then, for $x \in U$, the second derivative $D^2 f(x)$, when regarded as an element of* $\mathrm{Bil}(\mathbb{E}\times\mathbb{E},\mathbb{F})$, *is symmetric.*

Proof. We only need this in, iterations of, the elementary case $\mathbb{E} = \mathbb{F} = \mathbb{R}$. However, to give you a taste of the modifications of classical results needed to extend them to the infinite dimensional case, we give the following fairly straighforward adaption of the elementary proof, using the addendum above to the Mean Value Theorem.

Let $A(h,k) = f(x+h+k) - f(x+h) - f(x+k) + f(x)$ and $B(h) = f(x+h+k) - f(x+h)$. Then A is symmetric, $A(h,k) = B(h) - B(0)$ and $DB(h) = Df(x+h+k) - Df(x+h)$. Given ϵ we may choose δ such that $\| D^2f(x+h) - D^2f(x) \| < \epsilon$ whenever $\| h \| < \delta$. Now fix k with $\| k \| \leqslant \delta/2$ and restrict h to $\| h \| \leqslant \delta/2$. Then

$$\begin{aligned} &\| A(h,k) - (D^2f(x)\cdot k)\cdot h \| \\ &\leqslant \| B(h) - B(0) - DB(0)\cdot h \| \ + \ \| DB(0)\cdot h - (D^2f(x)\cdot k)\cdot h \| \ . \end{aligned} \tag{9.2}$$

But $\| DB(h) - D^2f(x+h)\cdot k \| = \| Df(x+h+k) - Df(x+h) - D^2f(x+h)\cdot k \| \leqslant 2\epsilon \|k\|$ by the addendum to the Mean Value Theorem and our hypothesis. (This is a somewhat terse indication of the proof, but readers should be able to fill in the details.) So in particular $\| DB(0) - D^2f(x)\cdot k \| \leqslant 2\epsilon \|k\|$ and, using the hypothesis again, $\| DB(h) - D^2f(x)\cdot k \| \leqslant 3\epsilon \|k\|$. Now, applying the extended Mean Value Theorem to the first term on the right hand side of (9.2), we get

$$\begin{aligned} \| A(h,k) - (D^2f(x)\cdot k)\cdot h \| &\leqslant \| h \| \sup_{\theta \in [0,1]} \| DB(\theta h) - DB(0) \| \ + 2\epsilon \|k\| \\ &\leqslant 7\epsilon \|k\| \, \|h\| \qquad \text{for} \quad \| k \| \, , \ \| h \| < \delta/2. \end{aligned}$$

Then the symmetry of $A(h,k)$ means that, for all ϵ, there is a δ such that

$$\| (D^2f(x)\cdot k)\cdot h - (D^2f(x)\cdot h)\cdot k \| \leqslant \epsilon \|h\| \, \|k\|$$

for all h, k with $\| h \| < \delta/2$ and $\| k \| < \delta/2$ and hence, by bilinearity, for all h and k. But ϵ is arbitrary so $(D^2f(x)\cdot k)\cdot h = (D^2f(x)\cdot h)\cdot k$. ∎

9.4. The Inverse Function Theorem.

As we saw in Chapter 1, this is the result that accounts for the very existence of differential manifolds.

Theorem 9.4.1. *Let U be open in the Banach space $\mathbb{E}$ and $f: U \to \mathbb{F}$ be of class C^1 with $Df(x)$ an isomorphism of $\mathbb{E}$ onto $\mathbb{F}$ for some $x \in U$. Then f is a local C^1-diffeomorphism at x.*

Proof. Since translation and linear isomorphisms are global, that is, everywhere local diffeomorphisms, we may assume $x = 0$, $f(0) = 0$, and $Df(0) =$id, the last two by replacing $f(x)$ by $Df(0)^{-1} \circ (f(x) - f(0))$ and using the chain rule. Now consider the related functions

$$g_y(x) = x + y - f(x) = y + g_0(x).$$

Then g_y is of class $\mathcal{C}^1$ and $g_0(0) = 0$, $Dg_0(0) = 0$. Since the derivative is continuous there exists $r > 0$ such that $\| x \| \leqslant r$ implies $\| Dg_0(x) \| \leqslant 1/2$. Then

$$\| g_y(x') - g_y(x) \| = \| g_0(x') - g_0(x) \| \leqslant \frac{1}{2}\|x' - x\|$$

by the Mean Value Theorem. So, if $\| y \| \leqslant r/2$, we have

$$\|g_y(x)\| = \| y + g_0(x) \| \leqslant \frac{r}{2} + \frac{\|x\|}{2} \leqslant r.$$

Thus, for $\| y \| \leqslant r/2$, g_y is a contraction mapping of the closed, and therefore complete, ball $\overline{B(0,r)}$ into itself. So there is a unique $x \in \overline{B(0,r)}$ such that $g_y(x) = x$, that is, such that $f(x) = y$. In particular $f(\overline{B(0,r)})$ contains the neighbourhood $B(0,r/2)$ of $f(0) = 0$ and $U_x = f^{-1}(B(0,r/2))$ is an open neighbourhood of x which f maps bijectively onto its image.

Choose δ such that $\overline{B(0,\delta)} \subseteq U_x$ then, since (in the finite-dimensional case, a more subtle argument being required for the infinite dimensional case) $\overline{B(0,\delta)}$ is compact and $\mathbb{F}$ is Hausdorff, $f|_{\overline{B(0,\delta)}}$ is a homeomorphism (see Lemma 9.1.6) and so too is $f|_{B(0,\delta)}$. Choosing δ sufficiently small that Df remains non-singular on $B(0,\delta)$ and reapplying the previous part at each point we see that $f(B(0,\delta))$ is also open. Thus f is a local homeomorphism on $B(0,\delta)$. ■

The following lemma is the basis of the proof that the inverse function is also differentiable.

Lemma 9.4.2. 1) *If $f: U \to V$ is a homeomorphism between open subsets of normed vector spaces and f is differentiable at $x \in U$, then f^{-1} is differentiable at $f(x)$ if and only if $Df(x)$ is an isomorphism.*

2) *If f in (1) is C^k on U, then f^{-1} is C^k on V if and only if $Df(x)$ is an isomorphism for all $x \in U$.*

Proof. The necessity in each case is obvious from the chain rule. For sufficiency in part (1), we may as before assume for convenience that $x = 0$, $f(x) = 0$ and $Df(0) =$ id. Then $f(x) = x + \eta(x)\,\|x\|$ where $\| \eta(x) \| \to 0$ as $\| x \| \to 0$. Hence

$$x = f(f^{-1}(x)) = f^{-1}(x) + \eta(f^{-1}(x))\|f^{-1}(x)\|. \tag{9.3}$$

Now f^{-1} continuous implies $f^{-1}(x) \to 0$ as $x \to 0$. So there exists δ such that $\| x \| < \delta$ implies $|\eta(f^{-1}(x))| < 1/2$. Hence, from (9.3), $\| x \| \geqslant \| f^{-1}(x) \|/2$.

Now we may rewrite (9.3) as $f^{-1}(x) = x - \eta(f^{-1}(x))\|f^{-1}(x)\| = x + \zeta(x)\|x\|$, where $\zeta(x)$ must be $-\eta(f^{-1}(x))\|f^{-1}(x)\|/\|x\|$ which does, as required, tend to 0 as x does, since, $\|f^{-1}(x)\|/\|x\| < 2$.

Now, for sufficiency in part (2), we immediately get that $g = f^{-1}$ is everywhere differentiable on V, and we also have from the chain rule that $Dg(f(x)) = (Df(x))^{-1}$. Or, writing $y = f(x)$, $Dg(y) = Df(g(y))^{-1}$. This is the composite of the three continuous functions g, Df and the map which sends a linear isomorphism to its inverse. Hence Dg is also continuous. $\mathcal{C}^r$ follows by induction. ∎

Addendum 9.4.3. Note that, for the Inverse Function Theorem, we have in fact proved that if f is $\mathcal{C}^r$, in addition to the other hypotheses, then so is f^{-1}. ∎

The Inverse Function Theorem has a number of relatives that are not strictly corollaries since, although they may all be deduced from the Inverse Function Theorem, it may equally be deduced from any one of them. These are discussed in Section 1.3. For the moment we emphasise only the Local Embedding Theorem, which is probably the one that most clearly shows the existence of submanifolds in Euclidean space. While the Inverse Function Theorem says that if the derivative is an isomorphism the map is a local diffeomorphism, the following theorem says that if the derivative is a linear isomorphism onto a subspace then locally, after a diffeomorphism of the image space, so too is the map.

Theorem 9.4.4. *Let U be open in $\mathbb{E}$ and $f: U \to \mathbb{F}_1 \times \mathbb{F}_2; x \mapsto (f_1(x), f_2(x))$ be a $\mathcal{C}^1$-function such that, without loss of generality, $0 \in U$ and $f(0) = 0$. If $Df_1(0) = D(\pi_1 \circ f)(0)$ is a linear isomorphism, then $\mathbb{E} \cong \mathbb{F}_1$ and there is a local diffeomorphism ϕ at $0 \in \mathbb{F}_1 \times \mathbb{F}_2$ such that $\phi \circ f = \iota_1$, the injection onto the first factor. Moreover, if ϕ is defined on the open neighbourhood V of 0 in $\mathbb{F}_1 \times \mathbb{F}_2$, then $\phi(V \cap f(U)) = \phi(V) \cap (\mathbb{F}_1 \times \{0\})$.*

Proof. Composing with $(Df_1)^{-1} \times \mathrm{id}$ we may further assume that $\mathbb{E} = \mathbb{F}_1$ and $Df_1(0) = \mathrm{id}$. Then

$$\psi: U \times \mathbb{F}_2 \longrightarrow \mathbb{F}_1 \times \mathbb{F}_2; \quad (x, y) \mapsto f(x) + (0, y) = (f_1(x),\, f_2(x) + y)$$

has $D\psi(0) = \begin{pmatrix} I & O \\ Df_2 & I \end{pmatrix}$, with inverse $\begin{pmatrix} I & O \\ -Df_2 & I \end{pmatrix}$. Let $\phi : V \to U \times \mathbb{F}_2$ be a local inverse to ψ at $(0,0)$ granted by the Inverse Function Theorem. Then, for all x in $f^{-1}(V)$, $\phi f(x) = \phi \circ \psi(x, 0) = (x, 0) = \iota_1(x)$. Thus, $\phi(V \cap f(U)) \subseteq \phi(V) \cap (U \times \{0\})$. Conversely, if $(x, 0) = \phi(x', y') \in \phi(V) \cap (\mathbb{F}_1 \times \{0\})$, then $x \in U$ since $\phi(V) \subseteq U \times \mathbb{F}_2$ and $f(x) = \psi(x, 0) = \psi \circ \phi(x', y') = (x', y') \in V \cap f(U)$. ■

Remark 9.4.5. It is the final sentence in this theorem that characterises the map as a local embedding rather than merely as an immersion. See Section 1.3.

9.5. Sard's Theorem.

Another fundamental result that plays an important rôle at various stages in the development of the theory of manifolds is Sard's Theorem. The following account is based on those of de Rham (1984) and Milnor (1965[2]). The Theorem states that, for a differentiable mapping, the set of images of points at which the map is singular is negligible in a sense that we now make precise.

Definition 9.5.1. We shall refer to a subset $C \subseteq \mathbb{R}^m$ that is a product of m closed intervals: $C = [a_1, b_1] \times [a_2, b_2] \times \cdots \times [a_m, b_m]$ as a *hyperrectangle* or, if all the intervals have the same length, a *hypercube*. Its volume is $\mathrm{vol}(C) = \prod_{i=1}^{m} (b_i - a_i)$.

Definition 9.5.2. A subset A of $\mathbb{R}^m$ has *measure* $\leqslant \epsilon$ if there exist hyperrectangles C_i, $i = 1, 2, , \cdots, k, \cdots$, possibly an infinite sequence of them, such that $A \subseteq \bigcup_{i=1}^{\infty} C_i$ and $\sum_{i=1}^{\infty} \mathrm{vol}(C_i) \leqslant \epsilon$. The set A has *measure zero* if it has measure $\leqslant \epsilon$ for all $\epsilon > 0$.

Lemma 9.5.3. *A countable union of sets of measure zero has measure zero.*

Proof. Let $A_1, \cdots, A_n, \cdots$ be sets of measure zero. For each A_i, choose hyperrectangles C_{ij}, $j = 1, 2, \cdots$, such that $A_i \subseteq \bigcup_{j=1}^{\infty} C_{ij}$ and $\sum_{j=1}^{\infty} \mathrm{vol}(C_{ij}) < \epsilon/2^i$. Then $A = \bigcup_{i=1}^{\infty} A_i \subseteq \bigcup_{i,j=1}^{\infty} C_{ij}$ and $\sum_{i,j=1}^{\infty} \mathrm{vol}(C_{ij}) < \sum_{i=1}^{\infty} \epsilon/2^i = \epsilon$. ■

Theorem 9.5.4. (Sard's Theorem.) *Let U be open in $\mathbb{R}^n$ and $f : U \to \mathbb{R}^m$ be a C^1-function. If A is the set $\{x \in U \mid \det(Jf(x)) = 0\}$ of singular points of f, then the set $f(A)$ of singular values has measure zero.*

Proof. We consider first the case $n = m$. Since U may be expressed as the union of a countable sequence of hyperrectangles we may, by the lemma, restrict attention to a single hyperrectangle contained in U which, after a Euclidean transformation of the domain U, we may take to be the unit hypercube $\mathbf{I}^m = [0,1]^m$ of diameter $\sqrt{m}$.

Given $\epsilon > 0$, the differentiability of f means that for every $x \in \mathbf{I}^m$ there is a $\delta(x)$ such that

$$\|f(y) - f(x) - Df(x) \cdot (y - x)\| < \epsilon \|y - x\| \tag{9.4}$$

whenever $\|y - x\| < \delta(x)$. The intersection $A \cap \mathbf{I}^m$ is closed and bounded and hence compact, so finitely many hypercubes C_i, $i = 1, \cdots, k$, centred on $y_i \in A \cap \mathbf{I}^m$ and of side $\delta(y_i)/\sqrt{m}$, cover $A \cap \mathbf{I}^m$.

Divide $\mathbf{I}^m$ regularly into p^m subcubes J_k, $k = 1, 2, \cdots, p^m$, of side length $1/p$ where $1/p$ is less than $\delta(y_i)/\{2\sqrt{m}\}$, $i = 1, \cdots, k$. Then, for each J_k that meets the singular set A, a common point $y \in J_k \cap A$ lies in some cube C_i and hence within $\delta(y_i)/2$ of y_i. Hence any other point y' of $J_k \cap A$ lie within $\delta(y_i)$ of y_i and so, by (9.4),

$$\|f(y') - f(y) - Df(y) \cdot (y' - y)\| \leqslant \epsilon \|y' - y\| \leqslant \frac{\epsilon\sqrt{m}}{p}.$$

Now, since $Df(y)$ is singular, the points $f(y) + Df(y) \cdot (y' - y)$ lie in an affine hyperplane through $f(y)$ and hence in a hypercube in the hyperplane, centred on $f(y)$, with side of length

$$2 \sup_{y' \in J_k \cap A} \|Df(y) \cdot (y' - y)\| \leqslant 2 \sup_{y' \in J_k} \|Df(y)\| \, \|y' - y\| \leqslant 2M\frac{\sqrt{m}}{p},$$

where $M = \sup_{x \in \mathbf{I}^m} \|Df(x)\|$ exists by the continuity of Df, and hence of its norm, on the compact set $\mathbf{I}^m$. Since $f(y')$ lies within $\epsilon\sqrt{m}/p$ of this $(m-1)$-dimensional hypercube of side $2M\sqrt{m}/p$, we see that the image $f(J_k \cap A)$ is contained in a hyperrectangle of volume

$$\frac{2\epsilon\sqrt{m}}{p}\left(\frac{2M\sqrt{m}}{p}\right)^{m-1} = (2\sqrt{m})^m M^{m-1}\frac{\epsilon}{p^m} = K\frac{\epsilon}{p^m},$$

where K is a constant independent of the subcube J_k that we are considering. Since there are p^m such subcubes, it follows that $f(A \cap \mathbf{I}^m)$ is contained in hyperrectangles of total volume $\leqslant K\epsilon$. Since ϵ is arbitrary $f(A \cap \mathbf{I}^m)$, and hence also $f(A)$, is of measure zero.

Note that this special case also covers the cases $n < m$ when every point is critical, since then the unit cube would only be divided into $p^n < p^m$ subcubes.

For the general case, we first look at the subset A_k of A of points at which all partial derivatives of order up to $k-1$ vanish. Then, by Taylor's Theorem and the compactness of $\mathbf{I}^n$, we can, as above, find a uniform bound M such that, for $y \in A_k \cap \mathbf{I}^n$ and $y' \in \mathbf{I}^n$,

$$\|f(y') - f(y)\| \leqslant M\|y' - y\|^k.$$

Hence, as for the case $n = m$, $f(A_k \cap \mathbf{I}^n)$ is contained in hypercubes of total volume

$$p^n \left(2M\left(\frac{\sqrt{m}}{p}\right)^k\right)^m = K\,p^{n-km}$$

which tends to zero as p, the number of subdivisions of each edge of $\mathbf{I}^n$, tends to infinity, provided $k > n/m$.

It remains to deal with $y \in A_k \setminus A_{k+1}$, for general k, which we do by induction on the dimension of the domain. Let ϕ be a $(k-1)$st order partial derivative of f such that, without loss of generality, $\frac{\partial \phi}{\partial x_1}(y) \neq 0$. Since $\phi(y') = 0$, for all $y' \in A_k$, the map

$$h : U \longrightarrow \mathbb{R}^n; \quad x \mapsto (\phi(x), x_2, \cdots, x_n)$$

is a diffeomorphism of some neighbourhood of V of y, where its derivative is an isomorphism, onto an open set V', carrying $A_k \cap V$ into the hyperplane $\mathrm{H} = \{x \in \mathbb{R}^n |\, x_1 = 0\}$. Let $\bar{g}$ be the restriction to $V' \cap \mathrm{H}$ of the composite $g = f \circ h^{-1} : V' \to \mathbb{R}^m$. Then the critical values $f(A_k \cap V) = \bar{g} \circ h(A_k \cap V) \subseteq \bar{g}(V' \cap \mathrm{H})$ lie in the set of critical values of $\bar{g}$ which, since $\bar{g}$ is defined on H, has measure zero by induction.

The argument above works for $k \geqslant 2$. For $y \in A \setminus A_2$ assume, without loss of generality, that h is defined as above with $\phi = f_i$, the ith component of f. Now the set A' of critical points of g is $h(V \cap A)$, so the set $g(A')$ of critical values of g is equal to the set $f(V \cap A)$ of critical values of $f|_V$. Note that g maps the hyperplane $\mathrm{H}_t = \{x \in \mathbb{R}^n |\, x_1 = t\}$ into the hyperplane $\mathrm{K}_t = \{x \in \mathbb{R}^m |\, x_i = t\}$ and that the critical points of $\bar{g} = g|_{\mathrm{H}_t}$ are just those of g that lie in H_t, since $\frac{\partial g_i}{\partial x_j} = \delta_{ij}$ and $\frac{\partial \bar{g}_k}{\partial x_j} = \frac{\partial g_k}{\partial x_j}$ for $k \neq i$ and $j > 1$. By induction the critical values of $\bar{g}$ have measure zero in K_t. Thus $f(V \cap A) \cap \mathrm{K}_t$ has $(m-1)$-dimensional measure zero for all t. By the theorem of Fubini (Royden, 1968) this means that $f(V \cap A)$ has m-dimensional measure zero. ■

Remarks 9.5.5. Of course we could start the induction at $n = 0$ rather than $n = m$. However the elementary nature of the latter case seems worthy of note particularly since it suffices for some of our applications. The case $k > n/m$ is also unnecessary for real analytic functions since such a function with all its derivatives zero must be identically zero.

9.6. Modules and Algebras.

The first non-analytic topic, well two topics if you insist, is just for convenience: it is useful to be able to talk about an *algebra*. This is a vector space on which there is also defined an associative, but not necessarily commutative, product which we shall denote by juxtaposition, which is related to the vector space structure as follows:

(*i*) $\lambda(vw) = (\lambda v)w = v(\lambda w)$;
(*ii*) $u(v + w) = uv + uw$;
(*iii*) $(v + w)u = vu + wu$;

where λ is a scalar and u, v, w are vectors.

In other words, the vector space V is an algebra if there is a bilinear mapping $V \times V \to V$ that, when interpreted as a product, is associative.

Examples of algebras with which the reader will already be familiar include the algebra of $n \times n$ matrices with entries in a particular field and the algebra of polynomials with coefficients in a field. The main type of algebra that will concern us is the algebra of mappings from a set X into the real line, or any other field. This is a vector space in a natural way, with the obvious definition of pointwise addition and scalar multiplication. Taking more general products in the image space gives us the structure of an algebra. As already mentioned in Example 9.3.3(5), the algebras that will mostly concern us are the algebras of smooth functions on smooth manifolds.

There is another important class of algebras which are not generally associative. These are the Lie algebras that we met in the chapter on Lie groups, and implicitly earlier. A Lie algebra structure is also defined on an underlying vector space, with the product structure usually denoted by $[\,,\,]$ rather than juxtaposition to emphasise that we are not assuming associativity.

Definition 9.6.1. A *Lie algebra* is a vector space L together with a product law that is a bilinear map

$$L \times L \longrightarrow L; \quad (v, w) \mapsto [v, w],$$

such that $[u, u] = 0$ and $[u, [v, w]] + [v, [w, u]] + [w, [u, v]] = 0$ for all u, v, w in L.

Remarks 9.6.2. The identity $[u, u] = 0$, together with linearity, implies that $[u, v] = -[v, u]$. The second identity is known as the *Jacobi Identity.* If we also assume associativity it implies that $[u, [v, w]] = 0$ for all u, v and w.

In Chapter 8 we saw that the left-invariant vector fields on a Lie group form a Lie algebra. Other, more basic, examples are:

1) the space $\mathbb{R}^3$ with the 'cross product' as operation;
2) the algebra of real $n \times n$ matrices with $[A, B] = AB - BA$;
3) any real vector space with all brackets set equal to zero.

A Lie algebra of type (3) is called commutative, or abelian. Lie subalgebras and Lie homomorphisms have the obvious definitions.

An important type of mapping of an algebra into itself is the following.

Definition 9.6.3. A *derivation* of an algebra A over the field $\mathbb{F}$ is an $\mathbb{F}$-linear mapping $\delta : A \to A$ such that for all $a, b \in A$, $\delta(ab) = a\,\delta(b) + b\,\delta(a)$.

As we explain in Chapter 2, it is sometimes helpful to think of a vector field on a manifold as a derivation of the algebra of smooth functions on that manifold.

It is also useful to be able to talk about a generalisation of a vector space, which is called a *module.* This is an abelian group with a scalar multiplication with analogous axioms to a vector space but where the scalars are drawn from a ring, rather than from a field. Thus $n \times n$ matrices with entries from a ring R form an R-module. Every abelian group is, in fact a $\mathbb{Z}$-module.

We can also similarly extend the definition of an algebra over a field as defined above, to algebras over rings, or even algebras over algebras! Then an *algebra over the ring* R is an R-module with an associative, bilinear product mapping, where bilinearity for modules means just what it meant for vector spaces: only the scalars are different.

9.7. Tensor Products.

Given two vector spaces, V and W, over the same field $\mathbb{F}$ their tensor product $V \otimes W$ is the vector space X together with a bilinear map $\otimes : V \times W \to X$ such that, for every bilinear map $\beta : V \times W \to Y$ there is a unique *linear* map $\lambda : X \to Y$ such that $\beta = \lambda \circ \otimes$. Naturally this definition assumes that such a space X and map $\otimes$ exist and are essentially unique. This is of course the case or we would not be using the definition!

We shall only require the tensor product of dual spaces for which you may check the following isomorphism:

$$V^* \otimes W^* \cong \mathrm{Bil}\,(V \times W\,; \mathbb{R})$$

and, more generally, the r-fold tensor product

$$V^* \otimes \cdots \otimes V^* = \mathrm{Multilin}\,(V \times \cdots \times V\,; \mathbb{R}).$$

9.8. Exterior Products.

Finally we shall require exterior products. The abstract definition we gave for the tensor product is generally expressed by saying that 'the tensor product is universal for bilinear mappings'. Similarly the pth exterior power of a vector space V is universal for skew-symmetric multilinear mappings. Note that we have to take the exterior product of a space with itself since the definition of skew-symmetry requires the arguments to be interchangeable.

What this means in detail is that, given a vector space V its pth exterior power $\Lambda^p(V)$ is the vector space X together with a skew-symmetric p-linear map $\wedge : V \times \cdots \times V \to X$ such that, for every skew-symmetric p-linear map $\beta : V \times \cdots \times V \to Y$ there is a unique *linear* map $\lambda : X \to Y$ such that $\beta = \lambda \circ \wedge$.

Again this definition assumes that such a space X and map $\wedge$ exist and are essentially unique and again this is the case. In fact the pth exterior product is a quotient of the p-fold tensor product, and we also have the following isomorphism:

$$\Lambda^p(V^*) \cong \mathrm{Alt}^p\,(V, \mathbb{R}),$$

where Alt^p is the space of alternating (skew-symmetric) p-linear maps from the p-fold cartesian product of V with itself into the scalars $\mathbb{R}$.

Care should be exercised with this unnatural and potentially confusing identification of what should be a quotient space with a subspace.

SOLUTIONS FOR THE EXERCISES

The exercises at the end of the chapters that correspond to the current Cambridge course are those intended for students to work at and then discuss with a supervisor. For the benefit of readers working independently we include in this chapter solutions for those exercises and occasional further comments.

Chapter 1

1.1. The real line with two origins, the quotient of two copies of the real line with coordinates s and t respectively by the equivalence relation $s \sim t \iff s = t \neq 0$, with the quotient topology is locally homeomorphic with $\mathbb{R}$ but not Hausdorff. If we give it uncountably many origins, it is not second countable. An uncountable set with the discrete topology is Hausdorff but is neither second countable nor locally homeomorphic with $\mathbb{R}^m$ for $m > 0$. More relevantly, the capital letter Y, being a subset of a Hausdorff space, is itself Hausdorff but not locally homeomorphic with $\mathbb{R}^m$ for any $m \geqslant 0$. A set of at least two elements with the indiscrete topology is neither Hausdorff nor locally homeomorphic with $\mathbb{R}^m$ for $m \geqslant 0$. However, since it only has two open sets, it is certainly second countable.

1.2. In other words, we are checking that the same manifold is arrived at by these two methods, that is, the two sets of charts are compatible. Given that all positive values of the map norm: $x \mapsto \|x\|$ are regular, there are many suitable decompositions of $\mathbb{R}^m$ possible to satisfy the hypotheses of Corollary 1.3.2. If, for a point v in the unit sphere $\mathbf{S}^{m-1}$, we take the tangent space $\tau_v(\mathbf{S}^{m-1})$ for our standard copy of $\mathbb{R}^{m-1}$, then the induced chart would be $u \mapsto u - (u \cdot v)v$ for u in a neighbourhood of v on the sphere. This has inverse

$$w \mapsto w + \sqrt{1 - \|w\|^2}\, v,$$

and we calculate that composition with either of the stereographic projections of Example 1.1.7 (2) is differentiable.

1.3. U open in M implies $U \cap U_\alpha$ open, which implies that $\phi_\alpha(U \cap U_\alpha)$ is open in V_α and hence in $\mathbb{R}^m$.

Conversely, $\phi_\alpha(U \cap U_\alpha)$ open in V_α implies that $U \cap U_\alpha$ is open in U_α and hence in M. Thus $U = \bigcup_\alpha (U \cap U_\alpha)$ is open in M.

1.4. The Möbius transformation

$$f(z) = \frac{az+b}{cz+d} \quad \text{is equal to} \quad \frac{b - ad/c}{cz+d} + \frac{a}{c}$$

if $c \neq 0$. So, in all cases, it is a composition of transformations of the forms

$$z \mapsto z + \alpha, \quad z \mapsto \alpha z \quad \text{and} \quad z \mapsto 1/z.$$

The first two of these are global analytic homeomorphisms on, say, the coordinate space $\phi_N(\mathbf{S}^n \setminus \{N\})$. After the coordinate transformation $z \mapsto \zeta = 1/z$ to the space $\phi_2(\mathbf{S}^n \setminus \{S\})$, they become

$$\zeta \mapsto \frac{\zeta}{1 + a\zeta} \quad \text{and} \quad \zeta \mapsto \frac{\zeta}{a},$$

respectively. Both have a removable singularity at the origin that may be removed by mapping $\zeta = 0$ to $\zeta = 0$, that is, N to N.

The inversion $z \mapsto 1/z$ is a complex analytic mapping of $\mathbb{C} \setminus \{0\}$ to itself, and transforms to the same map in the other coordinate space. However, if we extend it by mapping S to N and *vice versa* then, in terms of the chart ϕ_N about S and ϕ_2 about N, the extended map is represented by the identity, which is certainly a complex analytic homeomorphism. Similarly, interchanging the rôles of ϕ_N and ϕ_2, we see that the extended mapping is also analytic near N.

1.5. To see that $O(n)$ is a submanifold, consider the map

$$f : \mathbb{R}^{n^2} \longrightarrow \{\text{symmetric matrices}\} \cong \mathbb{R}^{n(n-1)/2}; \quad A \mapsto A^T A,$$

together with the path $A + tB$ passing through a point $A \in f^{-1}(I)$, where I is the $n \times n$ identity matrix, so that A is then orthogonal. Since $f(A + tB) = I + t(A^T B + B^T A) + t^2 B^T B$, the coefficient of t defines $Df(A) \cdot B$, the derivative of f at A evaluated on the 'vector' B, and, in order to show that I_n is a regular point, we need to show that this linear map is onto $\mathbb{R}^{n(n-1)/2}$. This follows from the fact that, if C is an arbitrary symmetric matrix, and we put $B = AC/2$, then

$$A^T B + B^T A = \frac{A^T AC}{2} + \frac{C^T A^T A}{2} = C,$$

since A is orthogonal and $C^T = C$, where the superscript T indicates the transposed matrix. Of course $SO(n)$ is just one of the components of $O(n)$, the one containing I.

There are, at least, two interesting ways of seeing the identification of $SO(3)$ with $\mathbb{RP}^3$. The first involves the stereographic projection of $\mathbf{S}^2$ onto its 'equatorial' plane, regarded as the complex plane: thus the point $(x, y, z) \in \mathbf{S}^2$ maps to $\zeta = \frac{x+iy}{1-z} \in \mathbb{C}$. Then an element of $SO(3)$, that is a rotation of $\mathbf{S}^2$, corresponds to a Möbius transformation of the form

$$\zeta \mapsto \frac{a\zeta + b}{-\bar{b}\zeta + \bar{a}}, \qquad \text{where } |a|^2 + |b|^2 = 1.$$

Conversely, any transformation of that form corresponds to a unique rotation of $\mathbf{S}^2$. The pairs (a, b) of complex numbers subject to $|a|^2 + |b|^2 = 1$ clearly form the 3-sphere $\mathbf{S}^3 \subseteq \mathbb{C}^2 \cong \mathbb{R}^4$ and distinct pairs (a, b) and (a', b') determine the same Möbius transformation if and only if $a' = -a$ and $b' = -b$. That is, if and only if they correspond to antipodal points on $\mathbf{S}^3$. However, the quotient of $\mathbf{S}^3$ by its antipodal map is $\mathbb{RP}^3$.

As an aside on the relations between the first members of the various natural sequences of Lie groups, we note that the matrices $\begin{pmatrix} a & b \\ -\bar{b} & \bar{a} \end{pmatrix}$ with $a, b \in \mathbb{C}$ and $|a|^2 + |b|^2 = 1$ form the Lie group $SU(2)$, comprising the matrices of determinant $+1$ preserving a Hermitian inner product on $\mathbb{C}^2$.

A second way of identifying $SO(3)$ with $\mathbb{RP}^3$ involves the quaternions. We recall that these are a 4-dimensional associative, but non-commutative, real algebra $\mathbb{H}$ with vector basis usually denoted $1, i, j, k$ and multiplication then determined by the relations

$$i^2 = j^2 = k^2 = -1, \qquad i\,j\,k = 1.$$

Within $\mathbb{H}$ we have the 3-dimensional subspace of *pure quaternions*:

$$\mathbb{H}^0 = \{ai + bj + ck \mid a, b, c \in \mathbb{R}\}.$$

The quaternions have a conjugation $q = a + bi + cj + dk \mapsto \bar{q} = a - bi - cj - dk$ which is related to the standard inner product, identifying $\mathbb{H}$ with $\mathbb{R}^4$, by

$(q, r) = \frac{1}{2}(q\bar{r} + r\bar{q}) = \sum q_i r_i$, where $q = q_1 + q_2 i + q_3 j + q_4 k$ and r is similar. The required identification arises from the fact that the unit quaternions $Q = \{q \in \mathbb{H} \mid \|q\| = 1\}$ form a submanifold identifiable with $\mathbf{S}^3$ and act as rotations on $\mathbb{H}^0 \cong \mathbb{R}^3$ by $q(h) = qhq$ for $q \in Q$ and $h \in \mathbb{H}^0$. That this is norm preserving is clear: that it has determinant 1 follows from the fact that Q is connected and contains 1 which acts as the identity. The kernel of the corresponding representation $\rho : Q \cong \mathbf{S}^3 \to SO(3)$ is $\{\pm 1\}$ so, once again, we get $\mathbb{R}\mathrm{P}^3 \cong \mathbf{S}^3/(\text{antipodal map}) \cong SO(3)$.

1.6. The conditions stated in the first sentence ensure that the quotient topology on M/G is Hausdorff and the fact that we have an action of G makes the quotient map $q : M \to M/G$ open and, hence, M/G second countable. Further, for a neighbourhood U as described, restricted to lie in the domain of a chart ϕ on M, $(q|_U)^{-1}$ is well-defined and we may take the chart $\phi \circ (q|_U)^{-1}$ on $q(U) \subseteq M/G$. If $\psi \circ (q|_V)^{-1}$ is another such with $q(U) \cap q(V) \neq \emptyset$, say $U \cap gV \neq \emptyset$, then the corresponding coordinate transformation is

$$\phi \circ (q|_U)^{-1} \circ (q|_V) \circ \psi^{-1} = \phi \circ (q|_U)^{-1} \circ (q|_{gV}) \circ g \circ \psi^{-1} = \phi \circ g \circ \psi^{-1},$$

which is smooth, where as usual we restrict the domains to ensure that the compositions are defined. See Figure 9.1 in which points of M in the same orbit lie above each other. Thus we have a differential structure on M/G for which q is expressed in terms of the related charts $\phi|_U$ and $\phi \circ (q|_U)^{-1}$ by the identity map.

Possible actions for the stated orbit spaces are as follows.

(*i*) $[0]$ acting as the identity and $[1]$ acting as the antipodal map.

(*ii*) For each $n \in \mathbb{Z}$, $n(t) = t + 2\pi n$. Then the orbit $[t]$ could be mapped to $e^{it} \in \mathbf{S}^1 \subseteq \mathbb{C}$ to provide the required diffeomorphism.

(*iii*) $$g : (x, y) \mapsto (x + 1, y), \quad h : (x, y) \mapsto (1 - x, 1 + y).$$
Note that the image in the Klein bottle of the vertical strip $1/3 < x < 2/3$ is a Möbius band, showing that the Klein bottle is also non-orientable.

1.7. On $\mathbf{S}^2$, without loss of generality, $z \neq 0$ and we may take (x, y) as local coordinates. Then, with standard coordinates on $\mathbb{R}^6$, the map, in coordinate form, is

$$(x, y) \mapsto (x^2, y^2, 1 - x^2 - y^2, \sqrt{2}y\sqrt{1 - x^2 - y^2}, \sqrt{2}x\sqrt{1 - x^2 - y^2}, \sqrt{2}xy),$$

of which the Jacobian is easily checked to have rank 2 everywhere for $x^2 + y^2 < 1$. Note that f factors through the quotient map $q : \mathbf{S}^2 \to \mathbb{R}\mathrm{P}^2$, $f = \bar{f} \circ q$ where

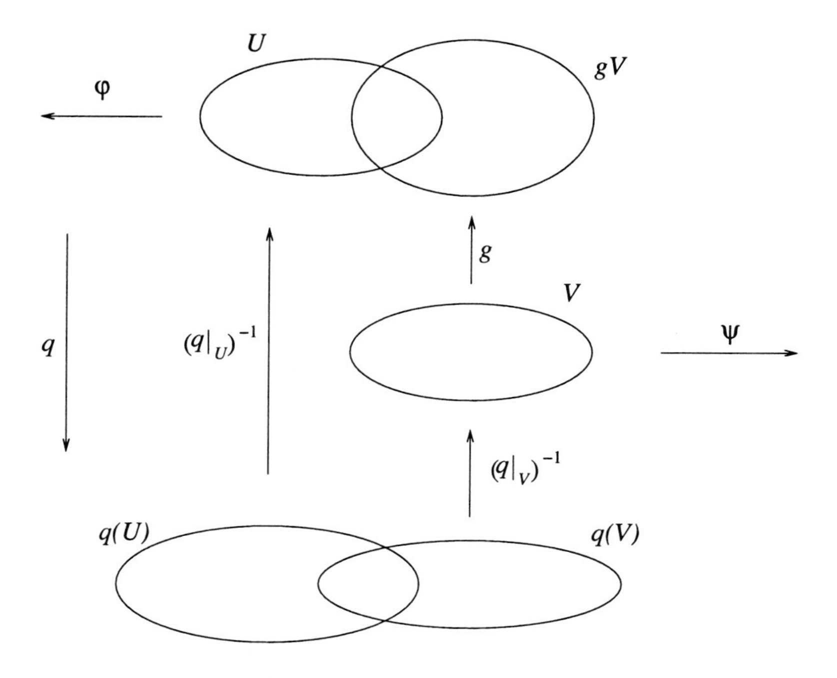

Figure 9.1.

$\bar{f}$ is differentiable and bijective from $\mathbb{R}\mathrm{P}^2$, which is compact, onto $f(\mathbf{S}^2) \subseteq \mathbb{R}^6$, which is Hausdorff. Thus $\bar{f}$ is an embedding of $\mathbb{R}\mathrm{P}^2$ in $\mathbb{R}^6$.

1.8. In both cases we use Lemma 1.3.12. A composite of injective immersions is certainly injective and of maximal rank. It is an open mapping, relative to its image since, for $f : M \to N$, $g : N \to P$ and U open in M, $f(U) = V \cap (f(M))$ where V open is N and $g(V) = W \cap (g(N))$, with W open in P, implies that $g \circ f(U) = g(V) \cap (g \circ f(M)) = W \cap (g \circ f(M))$.

Similarly, a product of homeomorphisms onto their images is a homeomorphism onto the product of the images, which is the image of the product, and a product of maps of ranks m_1 and m_2 has rank $m_1 + m_2$, the dimension of the product domain.

The application is proved by induction, starting from the embedding

$$\mathbf{S}^m \times \mathbb{R} \hookrightarrow \mathbb{R}^{m+1}; \quad (x, t) \mapsto e^t x.$$

Then

$$\mathbf{S}^{m_1} \times \cdots \times \mathbf{S}^{m_k} \times \mathbb{R} \hookrightarrow \mathbb{R}^{1+m_1+\cdots+m_k}$$

implies that

$$\begin{aligned}\mathbf{S}^{m_0} \times \mathbf{S}^{m_1} \times \cdots \times \mathbf{S}^{m_k} \times \mathbb{R} &\hookrightarrow \mathbf{S}^{m_0} \times \mathbb{R} \times \mathbb{R}^{m_1+\cdots+m_k} \\ &\hookrightarrow \mathbb{R}^{1+m_0+m_1+\cdots+m_k},\end{aligned}$$

showing the stronger result that M^m embeds with a trivial normal bundle, c.f. Definition 3.3.5 and Remark 7.4.2.

Chapter 2

2.1. Checking that $\mathcal{D}_p(M)$ is a vector space is routine.

A chart ϕ at p gives rise to an isomorphism $\tau_p(M) \cong \tau_0(\mathbb{R}^m)$ and also to a map $\mathcal{D}_p(M) \to \mathcal{D}_0(\mathbb{R}^m)$; $\delta \mapsto \delta \circ \phi^*$ which, again, is an isomorphism. Thus it suffices to work at $0 \in \mathbb{R}^m$.

There, we use the derivations $\partial_i(0) \in \mathcal{D}_0(\mathbb{R}^m)$ to define the linear (easy to check) map

$$\theta : \tau_0(\mathbb{R}^m) \longrightarrow \mathcal{D}_0(\mathbb{R}^m); \quad [\gamma] \mapsto \sum_{i=1}^{m} (\gamma'(0))_i \partial_i(0).$$

Thus, $\theta([\gamma])$ is the derivation

$$f \mapsto \sum_{i=1}^{m} (\gamma'(0))_i \frac{\partial f}{\partial x_i}(0) = (f \circ \gamma)'(0).$$

This is well-defined since $[\gamma] = [\delta] \Rightarrow \gamma'(0) = \delta'(0)$. The basis vector $[\delta_i]_0$ of $\tau_0(\mathbb{R}^m)$ maps to $\partial_i(0)$ so it suffices to show that these are linearly independent, which follows from $\left(\sum_{i=1}^{m} \lambda_i \partial_i(0)\right)(\pi_j) = \lambda_j$, and that they span $\mathcal{D}_0(\mathbb{R}^m)$. For the latter we show that, for an arbitrary derivation, $\delta = \sum_{i=1}^{m} \delta(\pi_i)\partial_i(0)$. This follows from the facts that, for any function f, $f(x) - f(0) = \sum_{i=1}^{m} \pi_i(x)\, g_i(x)$, where $g_i(x) = \int_0^1 \frac{\partial f}{\partial x_i}(tx)\, dt \in \mathcal{C}^\infty(\mathbb{R}^m)$ and that, if k_c denotes the function taking constant value c, then for $\delta \in \mathcal{D}_0(\mathbb{R}^m)$,

$$\delta(k_1) = \delta(k_1 \cdot k_1) = 1 \cdot \delta(k_1) + 1 \cdot \delta(k_1) = 2\delta(k_1).$$

So $\delta(k_1) = 0$ and, by the linearity of δ, $\delta(k_c) = 0$ for all c. Then, noting that $g_i(0) = \frac{\partial f}{\partial x_i}(0)$, we have

$$\left(\delta - \sum_{i=1}^m \delta(\pi_i)\partial_i(0)\right) f = \delta\left(f - \sum_{i=1}^m \frac{\partial f}{\partial x_i}(0)\,\pi_i\right) = \delta\left(\sum_{i=1}^m (g_i - g_i(0))\,\pi_i\right),$$

which is zero as required.

2.2. For part (i) see Example 2.3.7. For part (ii) see Theorem 3.2.3.

2.3. Convergence of the sequence follows from $\|A^n\| \leqslant \|A\|^n$ and, from the Jordan normal form (over $\mathbb{C}$) of the matrix, we also see that $\det(\exp(A)) = \exp(\mathrm{trace}(A))$: the eigenvalues of $\exp(A)$ are the exponentials of those of A. In fact, since sA and tA commute, $\gamma(s)\,\gamma(t) = \gamma(s+t)$ showing that γ is a 1-parameter subgroup. We discuss these in more detail in Chapter 8. In particular, $\gamma(s)\,\gamma(-s) = \gamma(0) = I$, so $\gamma(s)^{-1} = \gamma(-s)$. If A is skew-symmetric, $(\exp(A))^T = \exp(A^T) = \exp(-A) = (\exp(A))^{-1}$, so that $\exp(A)$ is in $O(m)$ and $\det(\exp(A)) = \exp(\mathrm{trace}(A)) = 1$ shows that it is in $SO(m)$.

Then $\gamma(s+h) - \gamma(s) = \gamma(s)\,(\gamma(h) - \gamma(0))$ shows that $\tau_g(SO(m)) = g\tau_I(SO(m))$ and $\lim_{h\to 0} \frac{\gamma(h)-\gamma(0)}{h} = A$ shows that $\tau_I(SO(m))$ is the space of skew-symmetric matrices, when we regard $SO(m)$ as a submanifold of $GL(m,\mathbb{R})$, whose tangent space at any point may be identified with the space of all $m \times m$ matrices.

2.4. Expressing M as a union of compact sets K_n such that $K_n \subset \mathrm{int}(K_{n+1})$ for each n as in the proof of Theorem 1.5.2, we may assume that $\mathrm{int}K_n$ is already immersed in $\mathbb{R}^{2m}$ or embedded in $\mathbb{R}^{2m+1}$. Then, choosing the charts used in the proof of Theorem 1.4.3 to lie either in $\mathrm{int}(K_n)$ or in $M \setminus K_{n-1}$ and using the already constructed immersion, resp. embedding, for the former, extend this to an immersion/embedding of K_{n+1} in some Euclidean space and project back to $\mathbb{R}^{2m}$, resp. $\mathbb{R}^{2m+1}$ as in the proof of Theorem 2.4.1. Since this projection does not move K_{n-1}, which is already in place, the inductive procedure converges to an immersion, resp. embedding, of M in the required dimensions.

Chapter 3

3.2. The existence of any allowable regular hyperbolic polygon may be proved implicitly by a continuity argument. However in this case there is the following more satisfying explicit construction.

One side of the octagon is an arc of a circle centred at a distance $\{(1+\sqrt{2})/2\}^{1/2}$ from the origin and of radius $r = \{2(1+\sqrt{2})\}^{-1/2}$. For the proof consider Figure 9.2 in which the vertex A of the octagon is placed on the real axis and OP passes through the mid-point of the (hyperbolic) side of the octagon through A. Then, since the interior angle of the octagon is $\pi/4$, $\alpha = \beta = \pi/8$, so $\gamma = \pi/4$ and hence $\delta = \pi/4$. Thus $OB = BA = AP = PQ = r$ and $OP = r(1+\sqrt{2})$ and then r may be calculated from $OP^2 = 1 + r^2$.

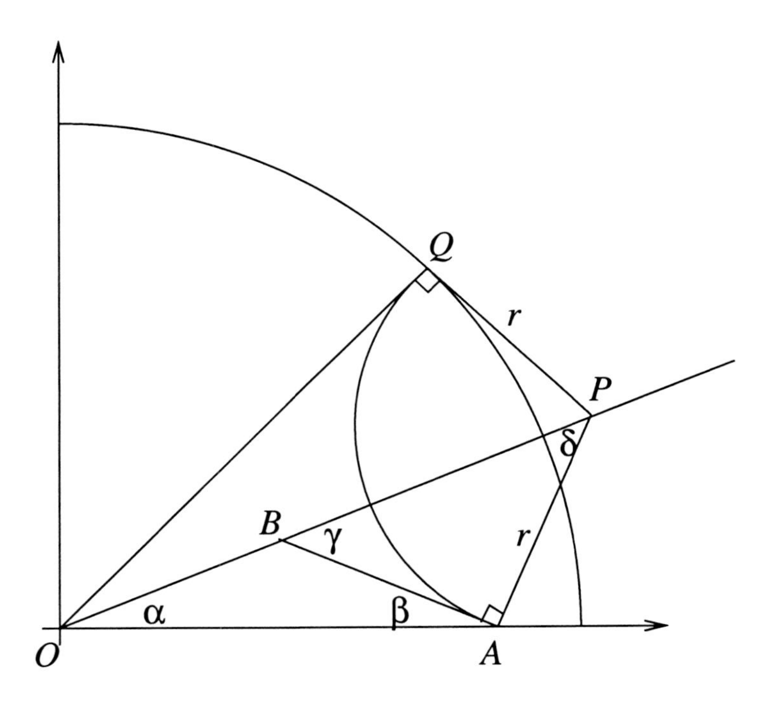

Figure 9.2.

Chapter 4

4.1. For $(i) \Rightarrow (ii)$, it suffices to observe that (i) implies that

$$\begin{aligned} 0 &= \mu(v_1 + v_2, v_1 + v_2, v_3, \cdots) \\ &= \mu(v_1, v_1, v_3, \cdots) + \mu(v_1, v_2, v_3, \cdots) \\ &\qquad\qquad + \mu(v_2, v_1, v_3, \cdots) + \mu(v_2, v_2, v_3, \cdots) \\ &= \mu(v_1, v_2, v_3, \cdots) + \mu(v_2, v_1, v_3, \cdots). \end{aligned}$$

For $(ii) \Rightarrow (iii)$, it follows from the fact that any permutation σ is the product of transpositions, the parity of the number of which depends only on σ.

For $(iii) \Rightarrow (i)$, simply transpose the equal arguments.

4.2.

$$\begin{aligned} (\alpha^*\omega)(e_1, \cdots, e_m) &= \omega(\alpha(e_1), \cdots, \alpha(e_m)) \\ &= \omega(a_{i_1 1}e_{i_1}, a_{i_2 2}e_{i_2}, \cdots, a_{i_m m}e_{i_m}) \\ &= a_{i_1 1}a_{i_2 2} \cdots a_{i_m m}\omega(e_{i_1}, \cdots, e_{i_m}) \\ &= \epsilon_\sigma a_{i_1 1} \cdots a_{i_m m}\omega(e_1, \cdots, e_m) \\ &= \det(\alpha)\, \omega(e_1, \cdots, e_m), \end{aligned}$$

where ϵ_σ is zero if any two of the i_k are equal and, otherwise, is the sign of the resulting permutation of 1, 2, ..., m. So, since $\omega = \lambda(e_1 \wedge \cdots \wedge e_m)$,

$$\alpha^* : \Lambda^m(V^*) \to \Lambda^m(V^*); \quad \omega \mapsto \det(\alpha)\, \omega.$$

4.3. If, without loss of generality, $\varphi_1 = \sum_{i=2}^{k} \lambda_i \varphi_i$, then

$$\varphi_1 \wedge \cdots \wedge \varphi_k = \left(\sum_{i=2}^{k} \lambda_i \varphi_i \right) \wedge \varphi_2 \wedge \cdots \wedge \varphi_k = 0.$$

Conversely, if the φ_i are linearly independent, we may extend them to a basis of V^* and take the dual basis $\{e_i \mid i = 1, \cdots, m\}$ of V. Then

$$(\varphi_1 \wedge \cdots \wedge \varphi_k)\, (e_1, \cdots, e_k) = 1$$

shows that $\varphi_1 \wedge \cdots \wedge \varphi_k \neq 0$.

Now, if $\varphi_i = a_{ij}\psi_j$, then

$$\begin{aligned}\varphi_1 \wedge \cdots \wedge \varphi_k &= a_{1j_1} a_{2j_2} \cdots a_{kj_k}\, \psi_{j_1} \wedge \cdots \wedge \psi_{j_k} \\ &= \sum_j \epsilon(j)\, a_{1j_1} \cdots a_{kj_k}\, \psi_1 \wedge \cdots \wedge \psi_k \\ &= \det(A)\, \psi_1 \wedge \cdots \wedge \psi_k,\end{aligned}$$

which is $\psi_1 \wedge \cdots \wedge \psi_k$ if $\det(A) = 1$, where we have written j for the permutation $1 \to j_1, 2 \to j_2, \cdots, k \to j_k$ and used the fact that the product $\psi_{j_1} \wedge \cdots \wedge \psi_{j_k} = 0$ if two factors are equal. Conversely, $\varphi_1 \wedge \cdots \wedge \varphi_k = \psi_1 \wedge \cdots \wedge \psi_k \neq 0$ implies that, for each i, $\varphi_i \wedge \psi_1 \wedge \cdots \wedge \psi_k = 0$ and so, as the ψ_j are linearly independent, $\varphi_i = \sum_{j_i=1}^{k} a_{ij_i}\psi_{j_i}$. Then, as above, $\det(A) = 1$.

4.4. $\dim(V) = 0$ or 1 implies that $\Lambda^2(V^*) = 0$, and $\dim(V) = 2$ implies that $\Lambda^2(V^*) = \langle e_1 \wedge e_2 \rangle$ and then the general 2-form $\lambda(e_1 \wedge e_2) = (\lambda e_1) \wedge e_2$ is decomposable. In dimension three, the general 2-form $a\, e_1 \wedge e_2 + b\, e_2 \wedge e_3 + c\, e_3 \wedge e_1$ may be written as

$$\begin{cases} (a\, e_1 - b\, e_3) \wedge (e_2 - \frac{c}{a} e_3) & \text{if } a \neq 0, \\ (b\, e_2 - c\, e_3) \wedge e_1 & \text{if } a = 0. \end{cases}$$

In dimension four, $\omega = a\, e_1 + b\, e_2 + c\, e_3 + d\, e_4$ and $\eta = \alpha\, e_1 + \beta\, e_2 + \gamma\, e_3 + \delta\, e_4$ with $\omega \wedge \eta = e_1 \wedge e_2 + e_3 \wedge e_4$ implies that

$$a\gamma = c\alpha,\ a\delta = d\alpha,\ b\gamma = c\beta,\ b\delta = d\beta,\ a\beta - b\alpha = 1 \text{ and } c\delta - d\gamma = 1.$$

Then $a = 0 \iff \alpha = 0$, etc., since $a = b = 0$ and $c = d = 0$ are impossible, and then, without loss of generality, $a \neq 0$, $\alpha \neq 0$. Then $a\gamma d\alpha = c\alpha a\delta$, so $c\delta - d\gamma = 0 \neq 1$.

4.5. $\Omega^2(\mathbb{R}^3)$, respectively $\Omega^2(\mathbb{R}^4)$, has

$$\begin{aligned} *(dx^1 \wedge dx^2) &= dx^3 && \text{resp.} && dx^3 \wedge dx^4 \\ *(dx^1 \wedge dx^3) &= -dx^2 && \text{resp.} && -\,dx^2 \wedge dx^4 \\ *(dx^2 \wedge dx^3) &= dx^1 && \text{resp.} && dx^1 \wedge dx^4. \end{aligned}$$

So $*\omega = a_{23}dx^1 - a_{13}dx^2 + a_{12}dx^3$, respectively (the same)$\wedge dx^4$.

Chapter 5

5.1. (*i*)

$$\begin{aligned} df(p)\,(e_j(p)) &= df(p)\,([\delta_j]_p) = (f\circ\delta_j)'(0) \\ &= \frac{\partial f}{\partial x_j}(p) = \left\langle \frac{\partial f}{\partial x_i}(p)\,e_i(p),\, e_j(p)\right\rangle. \end{aligned}$$

So $df(p)(Y) = \left\langle \frac{\partial f}{\partial x_i}(p)e_i(p),\, Y\right\rangle = \langle \mathrm{grad} f(p), Y\rangle$, for all $Y \in \tau_p(M)$, since this is true for basis vectors. Thus $\theta(\mathrm{grad} f) = df$.

(*ii*) Since $\mathrm{grad}\,\pi_j = \frac{\partial \pi_j}{\partial x_i} e_i = e_j$, we have $\theta(e_j) = d\pi_j = dx^j$. Then $\theta(X) = a_i\, dx^i$, so that $d(\theta(X)) = \frac{\partial a_i}{\partial x_j}(dx^j \wedge dx^i)$ and

$$*d\theta(X) = (\partial_2 a_3 - \partial_3 a_2)\, dx^1 + (\partial_3 a_1 - \partial_1 a_3)\, dx^2 + (\partial_1 a_2 - \partial_2 a_1)\, dx^3,$$

θ^{-1} of which is curl$X$.

Similarly, we find $\theta^{-1}(*(\theta(X)\wedge\theta(Y))) = X \times Y$.

5.2.

$$\begin{aligned} \omega(p)\,([\delta_1]_p) = a(x,y,z) &= \int_0^1 \frac{d}{dt}(t\,a(tx,ty,tz))\,dt \\ &= \int_0^1 t\,\{x\,\partial_1 a(tx,ty,tz) + y\,\partial_2 a(tx,ty,tz) + z\,\partial_3 a(tx,ty,tz)\}\,dt \\ &\quad + \int_0^1 a(tx,ty,tz)\,dt. \end{aligned}$$

But this is also $\frac{\partial f}{\partial x}(p) = df(p)\,(e_1(p)) = df(p)\,([\delta_1]_p)$. The result follows.

5.3. Observe that $d\omega = dx \wedge dy \wedge dz$ so that

$$\begin{aligned} \int_{\partial M} \iota^*(f^*(\omega)) = \int_M d(f^*(\omega)) &= \int_M f^*(dx\wedge dy\wedge dz) \\ &= \int_M dx\,dy\,dz = \mathrm{vol}(M), \end{aligned}$$

where $\iota : \partial M \hookrightarrow M$ is the inclusion, since we may use the standard Euclidean coordinates on M, in terms of which f is the identity.

When $M = B^3$ and $\partial M = \mathbf{S}^2$, we get $\int_{\mathbf{S}^2} \iota^*(f^*(\omega)) = \mathrm{vol}(B^3)$. However,

$$d\left(\frac{3\omega}{r^3}\right) = \left\{\frac{\partial}{\partial x}\left(\frac{x}{r^3}\right) + \frac{\partial}{\partial y}\left(\frac{y}{r^3}\right) + \frac{\partial}{\partial z}\left(\frac{z}{r^3}\right)\right\} dx \wedge dy \wedge dz$$
$$= \left\{\left(\frac{1}{r^3} - \frac{3x^2}{r^5}\right) + \left(\frac{1}{r^3} - \frac{3y^2}{r^5}\right) + \left(\frac{1}{r^3} - \frac{3z^2}{r^5}\right)\right\} dx \wedge dy \wedge dz,$$

which is zero since $\frac{\partial r}{\partial x} = \frac{x}{r}$ etc. Then, since r^{-3} is constant equal to 1 on $\mathbf{S}^2$, $\iota^*\left(f^*\left(\frac{1}{r^3}\omega\right)\right) = \iota^*(f^*(\omega))$ so that $d(\iota^*(f^*(\omega))) = d\iota^* f^*\left(\frac{\omega}{r^3}\right) = \iota^* \circ f^*\left(d\left(\frac{\omega}{r^3}\right)\right) = 0$. Thus $[\iota^*(f^*(\omega))] \in H^2(\mathbf{S}^2)$ and is non-zero, since $\iota^*(f^*(\omega)) = d\eta$ would imply

$$\int_{\mathbf{S}^2} \iota^*(f^*(\omega)) = \int_{\partial \mathbf{S}^2} \eta = 0$$

as $\partial \mathbf{S}^2 = \emptyset$.

5.4. If $X(p) = [\gamma]_p$, then $(Xf)(p) = f_*([\gamma]_p) = [f \circ \gamma]_p = (f \circ \gamma)'(0)$. Thus

$$(X(\lambda f + \mu g))(p) = ((\lambda f + \mu g) \circ \gamma)'(0)$$
$$= \lambda (f \circ \gamma)'(0) + \mu (g \circ \gamma)'(0) = (\lambda (Xf) + \mu (Xg))(p).$$

Also,

$$(X(fg))(p) = ((fg) \circ \gamma)'(0) = ((f \circ \gamma)(g \circ \gamma))'(0)$$
$$= (f \circ \gamma)(0)\,(g \circ \gamma)'(0) + (g \circ \gamma)(0)\,(f \circ \gamma)'(0)$$
$$= f(p)\,(Xg)(p) + g(p)\,(Xf)(p) = (f(Xg) + g(Xf))(p).$$

Conversely, for each derivation ξ of $\mathcal{C}^\infty(M)$ and each $p \in M$, $f \mapsto (\xi f)(p)$ is easily checked to be a 'derivation of $\mathcal{C}^\infty(M)$ at p'. So, by Exercise 2.1, it determines a unique vector $\xi(p) \in \tau_p(M)$ given locally on U by $\xi(p) = \xi(\pi_i)\, e_i(p)$. (See the proof in Section 10.2.) Then $\xi(\pi_i) \in \mathcal{C}^\infty(U)$ implies $\xi\big|_U$ is smooth.

The bracket $[X, Y]$ is certainly linear on $\mathcal{C}^\infty(M)$. The proof that it is a derivation is straightforward.

5.5. Clearly each side of the equation is additive in ω so, working locally, it suffices to consider $\omega = f\, dg$. Then $d\omega = df \wedge dg$ so

$$d\omega(X, Y) = df(X)\, dg(Y) - dg(X)\, df(Y) = (Xf)\,(Yg) - (Xg)\,(Yf).$$

However,

$$\begin{aligned}
&X(f\,dg(Y)) - Y(f\,dg(X)) - f\,dg([X,Y]) \\
=&X(f(Yg)) - Y(f(Xg)) - f(XYg - YXg) \\
=&(Xf)\,(Yg) + f(X(Yg)) - (Yf)\,(Xg) - f(Y(Xg)) \\
&\qquad - f(X(Yg)) + f(Y(Xg)) \\
=&(Xf)\,(Yg) - (Xg)\,(Yf).
\end{aligned}$$

Chapter 6

6.1. (*i*) Given a continuous map $f : M \to \mathbb{R}$ and $\epsilon > 0$, let $\delta(x)$ be choosen for each $x \in M$ such that $d(x,y) < \delta(x)$ implies $|f(x) - f(y)| < \epsilon$, where d is any distance function on M compatible with its topology (c.f. Corollary 1.4.4). Then let $\{\rho_x \mid x \in X\}$ be a (smooth) partition of unity supported on $\mathcal{U} = \{B(x, \delta(x) \mid x \in X\}$ for some subset X of M where $B(x,\delta(x)) = \{y \mid d(x,y) < \delta(x)\}$ and we may assume these supports are locally finite. Define $g : M \to \mathbb{R}$ by

$$g(y) = \sum_{x \in X} f(x)\rho_x(y).$$

Then, by the local finiteness of the supports and smoothness of the partition, g is both well-defined and smooth. However

$$\begin{aligned}
g(y) - f(y) &= \sum_{x \in X} \{f(x)\,\rho_x(y)\} - \left\{\sum_{x \in X} \rho_x(y)\right\} f(y) \\
&= \sum_{x \in X} \rho_x(y)\{f(x) - f(y)\}.
\end{aligned}$$

So, as $\rho_x(y) \geqslant 0$,

$$|g(y) - f(y)| \leqslant \sum_{x \in X} \rho_x(y)\,|f(x) - f(y)| < \epsilon$$

since either $\rho_x(y) = 0$ or $d(x,y) < \delta(x)$ and then $|f(x) - f(y)| < \epsilon$.

(*ii*) Let $\omega \in \Omega^k(N)$ be such that $d\omega = 0$. There is a cochain homotopy $h : \Omega^k(N) \to \Omega^{k-1}(N)$ such that $hd + dh = g^* \circ f^* - \mathrm{id}$, so $\omega = g^* \circ f^*(\omega) - dh(\omega)$.

But $f^*(\omega)$ in $\Omega^k(M)$ is closed so exact by hypothesis: $f^*(\omega) = d\eta$. Then $\omega = g^*(d\eta) - dh(\omega) = d(g^*(\eta) - h(\omega))$ is exact also.

(*iii*) M^m contractible implies $M \simeq \mathbb{R}^m$ so by (*ii*), as every m-form on M is closed, it must be exact. But this contradicts Corollary 5.5.6.

6.2. With $U = \mathbf{S}^n \setminus \{N\}$ and $V = \mathbf{S}^n \setminus \{S\}$, we have the chain of exact Mayer-Vietoris sequences of forms in Diagram 9.3 where, for convenience, we denote $\iota_{\mathbf{S}^n}^*$ by $\imath$ and $\iota_V^* - \iota_U^*$ by $\jmath$.

$$\begin{array}{ccccccccc}
0 & \to & \Omega^{k+1}(\mathbf{S}^n) & \xrightarrow{\imath} & \Omega^{k+1}(U \sqcup V) & & & & \\
 & & \uparrow d & [1] & \uparrow d & & & & \\
0 & \to & \Omega^{k}(\mathbf{S}^n) & \xrightarrow{\imath} & \Omega^{k}(U \sqcup V) & \xrightarrow{\jmath} & \Omega^{k}(U \cap V) & \to & 0 \\
 & & (4) & & (1) & & & & \\
 & & \uparrow d & [4] & \uparrow d & [2] & \uparrow d & & \\
0 & \to & \Omega^{k-1}(\mathbf{S}^n) & \xrightarrow{\imath} & \Omega^{k-1}(U \sqcup V) & \xrightarrow{\jmath} & \Omega^{k-1}(U \cap V) & \to & 0 \\
 & & & & (3) & & & & \\
 & & & & \uparrow d & [3] & \uparrow d & & \\
 & & & & \Omega^{k-2}(U \sqcup V) & \xrightarrow{\jmath} & \Omega^{k-2}(U \cap V) & \to & 0 \\
 & & & & & & (2) & &
\end{array}$$

Diagram 9.3.

Note that $U \cap V \simeq \mathbf{S}^{n-1}$ so, by Exercise 6.1(*ii*), we may assume every closed $(k-1)$-form on $U \cap V$ is exact. Also $U \simeq \{*\}$ and $V \simeq \{*\}$ so, by the Poincaré Lemma, every closed form in $\Omega^r(U) \oplus \Omega^r(V)$ for $r > 0$ is exact.

Assume $\omega \in \Omega^k(\mathbf{S}^n)$ is closed. Then $d(\imath(\omega)) = \imath(d\omega) = 0$ using commutative square [1]. Then $k \geqslant 1$ implies $\imath(\omega) = d\eta$ and

$$\begin{aligned}
d\jmath(\eta) &= \jmath(d\eta) && \text{by [2]} \\
&= \jmath \circ \imath(\omega) = 0 && \text{since } \ker(\jmath) \supseteq \operatorname{im}(\imath) \text{ at (1).}
\end{aligned}$$

Since $\jmath(\eta)$ is closed in $\Omega^{k-1}(U \cap V)$ it is exact: $\jmath(\eta) = d\zeta$ and $\jmath$ surjects (ker $\subseteq$ im($\jmath$) at (2)) so $\zeta = \jmath(\xi)$. Then

$$\begin{aligned}\jmath(d\xi) &= d(\jmath(\xi)) \qquad \text{by [3]}\\ &= d\zeta = \jmath(\eta)\end{aligned}$$

so $\jmath(\eta - d\xi) = 0$. Hence (ker $\subseteq$ im at (3)) $\eta - d\xi = \imath(\tau)$ for $\tau \in \Omega^{k-1}(\mathbf{S}^n)$. Then

$$\begin{aligned}\imath(d\tau) &= d(\imath(\tau)) \qquad \text{by [4]}\\ &= d\eta - d^2\xi = d\eta = \imath(\omega).\end{aligned}$$

So $\imath(d\tau - \omega) = 0$. But $\imath$ injects (im $\supseteq$ ker at (4)), so $d\tau = \omega$ as required.

The *proof* does not require $1 < k < n$, though it does require $k > 0$, but the *hypothesis* would fail if $k = 1$ or $k = n$. For then $\jmath(\eta) \in \Omega^{k-1}(\mathbf{S}^{n-1})$ could not be exact unless it were zero. However, if $k = 1$ and $n > 1$, then $\mathbf{S}^{n-1}$ is connected and so $d(\jmath(\eta)) = 0$, for $\jmath(\eta) \in \Omega^0(U \cap V)$, implies that $\jmath(\eta)$ is the constant function. Thus $\eta\big|_U$ and $\eta\big|_V$ differ by a constant c on $U \cap V$. Then we can define a function τ on $\mathbf{S}^n$ such that $\imath \circ \tau = (\eta, \eta + c)$ and then

$$\begin{aligned}\imath(d\tau) = d(\imath(\tau)) &= d\eta \qquad \text{since } dc = 0\\ &= \imath(\omega) \qquad \text{as before.}\end{aligned}$$

6.3. Given the detailed diagram chase that we have described in the text and in the above solution, the reader should not have any difficulty carrying out these remaining two examples.

6.4. Take $M = U \cup V$ where $U = M \setminus \{f(0)\} \cong \mathring{M}$ and $V = f(\mathbb{R}^m) \cong \mathbb{R}^m$ so that $U \cap V \simeq \mathbf{S}^{m-1}$. Then, for $1 < k < m - 1$, the Mayer-Vietoris sequence in cohomology gives

$$H^{k-1}(\mathbf{S}^{m-1}) = 0 \longrightarrow H^k(M) \longrightarrow H^k(\mathring{M}) \oplus H^k(\mathbb{R}^m) \longrightarrow H^k(\mathbf{S}^{m-1}) = 0$$

so that $H^k(M) \cong H^k(\mathring{M})$ in that range. Then M connected and $m > 1$ together imply that $\mathring{M}$ is connected and so $H^0(\mathring{M}) \cong \mathbb{R}$ and this, together with the datum that $H^m(\mathring{M}) = 0$, allows us to complete the special cases $k = 1$ and $k = m - 1$: the proof differs slightly but the result is the same, $H^k(\mathring{M}) \cong H^k(M)$ here too.

6.5. (*i*) A volume form on $M\#M$ restricts to one on $\mathring{M}$ and that, together with one on $f(\mathbb{R}^m)$, gives rise to one on M (possibly changing the sign of one of them) since the intersection of their domains is connected. Similarly volume forms on M and N give rise to one on $M\#N$ since, again, $\mathring{M} \cap \mathring{N}$ is connected.

(*ii*) Use the Mayer-Vietoris sequence determined by $M\#N = \mathring{M} \cup \mathring{N}$ with $\mathring{M} \cap \mathring{N} \simeq \mathbf{S}^{m-1}$. $H^0(M\#N) \cong \mathbb{R}$ since $M\#N$ is connected, $H^0(\mathring{M}) + H^0(\mathring{N}) \to H^0(\mathbf{S}^{m-1})$ clearly surjects and, for $1 < k < m$, $H^{k-1}(\mathbf{S}^{m-1}) = 0$. By (*i*) and $H^m(\mathring{M}) = 0 = H^m(\mathring{N})$, the segment

$$H^{m-1}(\mathbf{S}^{m-1}) \cong \mathbb{R} \longrightarrow H^m(M\#N) \cong \mathbb{R} \longrightarrow H^m(\mathring{M}) \oplus H^m(\mathring{N}) = 0$$

shows that the first map is an isomorphism, so its predecessor $H^{m-1}(\mathring{M}) \oplus H^{m-1}(\mathring{N}) \to H^{m-1}(\mathbf{S}^{m-1})$ is zero. Thus, for $0 < k < m$, we get

$$H^{m-1}(\mathbf{S}^{m-1}) \xrightarrow{0} H^k(M\#N) \longrightarrow H^k(\mathring{M}) \oplus H^k(\mathring{N}) \xrightarrow{0} H^k(\mathbf{S}^{m-1}),$$

so that $H^k(M\#N) = H^k(M) \oplus H^k(N)$.

Chapter 7

7.1. Writing α for the antipodal map and P for the point $(1, 0, \cdots, 0)$, let ρ be rotation through π of the $(1, m)$-plane with other axess fixed, so that $\rho \circ \alpha(P) = P$. The rotation ρ is homotopic to the identity so $\rho \circ \alpha$ is homotopic to α. The tangent space to $\mathbf{S}^{m-1}$ is spanned by vectors parallel to the $(2, 3, \cdots, m)$-axes of which all but the last are reversed by $\rho \circ \alpha$. Hence $\det(J(\rho \circ \alpha))(P) = (-1)^{m-2}$.

Let $\mathbf{S}^{m-1} \xrightarrow{q} \mathbf{S}^{m-1} / \alpha = \mathbb{R}\mathrm{P}^{m-1}$. If $\mathbb{R}\mathrm{P}^{m-1}$ is orientable, it has a never-zero volume form ω and, since q is a local diffeomorphism, $q^*(\omega)$ is also never-zero on $\mathbf{S}^{m-1}$. However, $q \circ \alpha = q$ so $\alpha^*(q^*(\omega)) = q^*(\omega)$, but for m odd α has odd degree so $\alpha^*(q^*(\omega)) = -q^*(\omega)$, a contradiction. Thus, $\mathbb{R}\mathrm{P}^n$ is non-orientable for $n = m - 1$ even.

7.2. Let $I_\alpha = \{e^{i\theta} \mid -\alpha < \theta < \alpha\} \subseteq \mathbf{S}^1$ with $\alpha < \pi$. Then I_α is the diffeomorphic image under θ_n of the n intervals $I_k = \{e^{i\theta} | \frac{-\alpha+2k\pi}{n} < \theta < \frac{\alpha+2k\pi}{n}\}$, $k = 0, 1, \cdots, n-1$. Moreover no other points map into I_α. This establishes that $\deg(\theta_n) = n$, since $\theta_n\big|_{I_k}$ is orientation preserving for each k.

Re-parameterised as a map on $\mathbf{S}^1$, rather than [0, 1] with $\gamma(0) = \gamma(1)$, the map γ becomes $\bar{\gamma}(e^{2\pi i t}) = a + r(t)\ \exp(i\theta(t))$. Then its winding number is the degree of the map

$$z = e^{2\pi i t} \mapsto \frac{r(t)\ \exp(i\theta(t))}{\|r(t)\ \exp(i\theta(t))\|} = \exp(i\theta(t)).$$

Since $\theta(0) = 0$ and $\theta(1) = 2\pi n$, $H_s(t) = (1-s)\theta(t) + s\,2\pi nt$ is a homotopy between θ and the map $\phi(t) = 2\pi nt$, inducing a homotopy between the above map and the map $z \mapsto z^n$. Since this has degree n, γ has winding number n about a.

7.3. Letting $z_j = x_j + iy_j$, the Hopf map is

$$(x_1, y_1, x_2, y_2) \overset{h}{\mapsto} \left(\frac{x_1x_2 + y_1y_2}{x_2^2 + y_2^2}, \frac{y_1x_2 - y_2x_1}{x_2^2 + y_2^2}\right) = (h_1, h_2)$$

restricted to $x_1^2 + y_1^2 + x_2^2 + y_2^2 = 1$. The group $SL(2, \mathbb{C})$, comprising 2×2 complex matrices of determinant 1, acts as a group of diffeomorphisms on $\mathbf{S}^3 \subseteq \mathbb{C}^2$ and as Möbius transformations on $\mathbf{S}^2$ regarded as $\mathbb{C} \cup \{\infty\}$. Since h is equivariant with respect to these actions, that is, $h\left(A\begin{pmatrix} z_1 \\ z_2 \end{pmatrix}\right) = f_A(z_1/z_2)$ where $f_A : z \mapsto \frac{az+b}{cz+d}$ is the Möbius transformation corresponding to the matrix

$$A = \begin{pmatrix} a & b \\ c & d \end{pmatrix} \in SL(2, \mathbb{C}),$$

and since the Möbius group is triply transitive on $\mathbb{C} \cup \{\infty\}$, it suffices to check that the origin is a regular value and to find the linking number between $h^{-1}(0)$ and $h^{-1}(\infty)$. Now $h^{-1}(0)$ is the circle $x_2^2 + y_2^2 = 1$, $x_1 = 0 = y_1$. At points of this circle, the submatrix

$$\begin{pmatrix} \frac{\partial h_1}{\partial x_1} & \frac{\partial h_1}{\partial y_1} \\ \frac{\partial h_2}{\partial x_1} & \frac{\partial h_2}{\partial y_1} \end{pmatrix}$$

of the Jacobian matrix of h is non-singular. Thus the origin is indeed a regular value.

One way to project $\mathbf{S}^3 \setminus \{\mathrm{pt}\}$ onto $\mathbb{R}^3$ is to copy the projection of $\mathbf{S}^2$ onto $\mathbb{C} \cup \{\infty\}$. Then, for example, (x_1, y_1, x_2, y_2) projects to $\left(\frac{y_1}{1-x_1}, \frac{x_2}{1-x_1}, \frac{y_2}{1-x_1}\right)$ with the point $(1, 0, 0, 0)$ mapping to the point '∞' that compactifies $\mathbb{R}^3$. Then the circle $h^{-1}(0)$ projects to the circle $\{(0, y, z) \mid y^2 + z^2 = 1\}$ and $h^{-1}(\infty)$ is the circle $x_1^2 + y_1^2 = 1$, $x_2 = 0 = y_2$, which projects to the x-axis $\{(x, 0, 0) \mid x \in \mathbb{R}\}$. One way to see that these have linking number ± 1, depending on their orientations, is to fix a point P on the x-axis. Then the unit vectors pointing towards points of the circle in the y-z plane form an 'umbrella' of rays. As P varies along the x-axis these rays trace out all possible directions in $\mathbb{R}^3$, so all points of $\mathbf{S}^2$, except for at most two, with none repeated. This is enough to ensure that the map that determines the linking number has degree ± 1.

7.4. Taking $\mathbf{S}^2 = \mathbb{C} \cup \{\infty\}$, the map $z \mapsto z^n$ has degree n. This is in fact the 'suspension' of the map θ_n in Exercise 7.2 and the construction generalises: assume Θ_n is defined on $\mathbf{S}^{m-1}$ with degree n; let the m-ball B^m have generalised polar coordinates (r, ϕ) where $r \in [0, 1]$ and $\phi \in \mathbf{S}^{m-1}$; define $\tilde{\Theta}_n$ on B^m by

$$(r, \phi) \mapsto (r, \Theta_n(\phi));$$

regard $\mathbf{S}^m$ as the union of two copies of B^m with their boundaries identified and hence extend Θ_n to $\tilde{\Theta}_n$ on $\mathbf{S}^m$. It is not difficult to check that $\tilde{\Theta}_n$ has the same degree as Θ_n: n disjoint $(m-1)$-balls that have the same image under Θ_n extend, on multiplying by a radial interval, to n disjoint m-balls having the same image under $\tilde{\Theta}_n$.

The above argument works for all (strictly) positive degrees. For negative degrees, compose with a map of degree -1, for example, a reflection in a hyperplane. A constant map has degree zero.

7.6. Use, for example, the fact that $\mathbf{S}^{2k-1}$ is the unit sphere in $\mathbb{C}P^k$.

Chapter 8

8.1. Both parts may be solved using the adjoint representation of a Lie group G: see Bröcker and tom Dieck (1985, pages 18-20). Ad:$G \longrightarrow$Aut(LG) takes the group element g to the derivative of the inner automorphism $x \mapsto gxg^{-1}$ at the identity. In turn 'Ad' induces a homomorphism of Lie algebras, written 'ad', such that $[X, Y] = \mathrm{ad}(X)\,Y$. Since inner automorphisms are trivial for abelian groups assertion (i) ($[X, Y] = 0$) follows immediately.

For (ii) let the Lie algebra elements X and Y be associated with the one-parameter subgroups

$$\phi^X(s) = 1 + sX \pmod{s^2}$$
$$\phi^Y(t) = 1 + tY \pmod{t^2}$$

and write $$\mathrm{ad}(X)Y = \left.\frac{\partial}{\partial s}\right|_0 \left.\frac{\partial}{\partial t}\right|_0 a(s, t)$$

with $a(s, t) = \phi^X(s)\phi^Y(t)\phi^X(-s)$. Thus

$$a(s,t) = (1 + sX)(1 + tY)(1 - sX) = 1 + tY + st(XY - YX)\mathrm{mod}(s^2, t^2),$$

which, on differentiating, gives $[X, Y] = XY - YX$ or (ii).

8.2. The fact that $K \cap yH = \emptyset$ imples that $Kh \cap yH = \emptyset$ for all $h \in H$. Hence KH is a neighbourhood of xH disjoint from yH. To prove closure consider $f : K \times G \longrightarrow G : (k, g) \mapsto k^{-1}g$. The element x belongs to KH iff $k^{-1}x \in H$ for some k. Therefore $KH = \mathrm{pr}_2(f^{-1}H)$ and $f^{-1}H$ is closed. Since K is compact, projection from $K \times G$ onto G is a proper closed map and KH is closed.

8.3. The map $q \mapsto aqb^{-1}$, where $a, b \in SU(2)$, associates an orientation and distance preserving linear map of $\mathbb{R}^4$ onto itself with each pair (a, b). This map is 2–1 with kernel $\{(I_2, I_2), (-I_2, -I_2)\}$. The fundamental group of $SO(4)$ is of order 2, the two-leaved universal cover is denoted Spin(4) and our map lifts to it. Compare the discussion of $SO(3)$ on pages 187-8 above.

GUIDE TO THE LITERATURE

Having reached the end of this elementary text, it seems reasonable to suggest further reading, particularly as the topics we have covered are, to some extent, dictated by the requirements of one university syllabus. One obvious alternative text is the book by Th. Bröcker and K. Jänich (1982), the disappearance of which from the bookstores provided one incentive to write our own replacement. With some shifts in emphasis their first six sections are similar to our Chapters 1 and 2, together with Sard's Theorem. Besides the elegant diagrams their careful initial treatment of tangent vectors is particularly to be recommended. The later sections of the book (Sections 7-14) are concerned with two topics essential to any second course in differential topology: smooth isotopy extension and transversality. For the first of these one shows that, at least for compact manifolds, the deformation of one submanifold onto another can be covered by a 1-parameter family of diffeomorphisms of the ambient manifold. For transversality we first need the definition: let $N^n \hookrightarrow M^m$ and $f : V^v \to M^m$ be differentiable. Then f is *transverse* to N if for each $x \in V$ with $f(x) = y \in N$, $f_*(\tau_*(V)) + \tau_y(N) = \tau_y(M)$. Neglecting the pedant's case when $f(V) \cap N = \emptyset$, a necessary dimensional condition is $v \geqslant m - n$. A consequence of transversality is that $f^{-1}(N) = W^w \subseteq V^v$ with equality of codimensions, that is, $v - w = m - n$. In its simplest form the transversality theorem states that an arbitrary differentiable map $f : V \to M$ can be approximated as closely as we please by a map transverse to N.

The second text immediately relevant to our own is Madsen and Tornehave, (1997), the first part of which concentrates on de Rham cohomology, a topic which has received no mention in Bröcker and Jänich (1982). Some knowledge of the differential geometry of surfaces is a useful pre-requisite, particularly as the second half of the book is devoted to the differential-geometric construction of characteristic classes of vector bundles. In the complex case these are elements in the even-dimensional cohomology of the base which carry information for an n-dimensional bundle generalising the $(1-1)$-correspondence $\mathrm{Vect}^1_{\mathbb{C}}(B) \longleftrightarrow H^2(B, \mathbb{Z})$. Like us Madsen and Tornehave include a section on the Poincaré-Hopf Theorem. Assuming some familiarity with the notion of curvature K, they also include a proof of the Gauss-Bonnet formula for a surface F embedded in $\mathbb{R}^3$, viz.

$$\frac{1}{2\pi}\int_F K\mathrm{vol}(F) = \chi(F).$$

This belongs to the same circle of ideas. Counting critical values of the Gauss map g for F shows that twice its degree equals $\chi(F)$. If we then define the curvature at the point x by $K(x) = \det(g_*(x))$ and note that $K\mathrm{vol}(F) = g^*(\mathrm{vol}(S^2))$, integration shows that

$$\int_F K\mathrm{vol}(F) = (\deg g)\int_{S^2}\mathrm{vol}(S^2) = 4\pi(\deg g).$$

They also prove Poincaré duality for differential manifolds between homology and cohomology with real coefficients. The result should be plausible to the reader who has done our extended exercise on the de Rham isomorphism. The essential point is that the manifold M^m admits a 'dual' handle decomposition, in which we add the handles in reverse order with the rôles of m and $m-k$ in the proof of Lemma 7.6.4 interchanged.

Another, in fact the first, text that uses differential forms to motivate and explore more general concepts is Bott and Tu (1982). However, as their title indicates, their scope is broader and they go much deeper than either Madsen and Tornehave (1997) or ourselves. For example they give a careful account of forms, and the resulting cohomology, with compact support, as well as introducing presheaves into the discussion. The overlap with our text is, very strictly, contained in their first two chapters after which they have two further substantial chapters, one on spectral sequences together with a wide range of applications and one on characteristic classes. The latter includes sections on flag manifolds, Grassmannians and universal bundles, so goes well beyond the brief introduction to the subject in Madsen and Tornehave (1997). Thus Bott and Tu (1982) is very much a secondary text.

The book of Lawrence Conlon (2001) could serve as required reading for a second course on differential topology. Leaving aside the final chapters on Riemannian geometry, the topics covered are similar to ours, but treated at greater length and developed further. An example is provided by the topic of *foliations*. Let us start with the global flow $\phi : \mathbb{R}\times M \to M$ associated with a vector field on M; the orbits of the $\mathbb{R}$-action cover M and define the leaves a 1-dimensional foliation. Defining a k-parameter family of diffeomorphisms to be a differentiable $\mathbb{R}^k$-action on M we similarly obtain a foliation in which the leaves are k-dimensional. Another example is provided by a fibration $E = \bigcup_{b \in B} F_b$, in which the leaves have the special property of being all diffeomorphic. In the same way that vector fields correspond to flows, integrable distributions (subbundles of τM) correspond to foliations. Let $\mathcal{D}$ be such a k-plane subbundle of the tangent bundle $\tau(M)$; then $\mathcal{D}$ is integrable, that is, at least locally tangent to a k-dimensional submanifold provided that the space of sections taking values in $\mathcal{D}$ is closed with respect to Lie brackets. Such a distribution is said to be *involutive* or to satisfy the Frobenius condition. Foliations were a fashionable area of research in the 1970s – as with submanifolds one can define a normal bundle to the leaves with fibre dimension equal to $\dim(M) - k$. There is a substantial literature on foliations with geometric (complex, Riemannian, or symplectic) structures in the normal direction, which alone suggests that the subject is ripe for a comeback.

Conlon's book also contains an honest appendix on ordinary differential equations. Honest in this context means that attention is paid to the smoothness of solution curves as the initial condition is varied. It is possible to give an elegant proof of this, provided one allows infinite dimensional manifolds, that is, manifolds in which the local coordinate charts are modeled on Hilbert or Banach space. The theory is developed in this degree of generality by S. Lang in updated versions of a book first published in 1962: see Lang (1995) for the latest. There is no doubt that Lang's book is extremely elegant; here is to be found for example an early use of sprays to construct tubular neighbourhoods. But like Conlon (2001) and Bott and Tu (1982) it is best left to a second course. We give another example of Lang's elegance: like Conlon he includes Frobenius' theorem on integrable distributions. Now consider the problem of characterising Lie subgroups of the Lie group G (Theorem 8.1.16). Let $\mathfrak{h}$ be a Lie subalgebra of $\mathfrak{g} = L(G)$, and let $\mathcal{D}$ be the corresponding left invariant subbundle of $\tau(G)$ obtained by translation. Then $\mathcal{D}$ satisfies the Frobenius condition and hence is integrable, giving rise to a maximal connected submanifold H of G passing through the identity element $e \in G$. Then H is a closed subgroup of G corresponding to the original subalgebra $\mathfrak{h}$.

Lang's book also suggests a further area for research. The $\mathcal{C}^\infty$ functions defined on an open subset of $\mathbb{R}^n$ do not form a Banach but a Fréchet space (with countably many (semi) norms rather than one). The implicit function theorem and the local existence theorem for ordinary differential equations are no longer true, and in order to obtain similar results one needs more sophisticated methods. These were developed by John Nash as part of his proof of the

Riemannian embedding theorem referred to in the introduction. For a survey of what was known in the 1980s see Hamilton (1982). To illustrate the importance of this area one has only to point out that the group of diffeomorphisms of a compact manifold Diff(M) has the structure of a Fréchet Lie group: see Hamilton (1982), page 98.

Consideration of function spaces brings us round to the last of our alternative texts, Hirsch (1976). This is perhaps the most idiosyncratic of the books we have considered, proving the transversality theorem for example, before introducing vector bundles and tubular neighbourhoods. It is worth pointing out that Hirsch's treatment of transversality is very general; not only does he consider the density of transverse maps in $\mathcal{C}^\infty(V^v, M^m)$, an application of the Baire Category Theorem, but he also allows N^n to be a submanifold of the *jet space* $J^r(V, M)$ rather than of M alone: see Hirsch (1976), Section 2.4 for the definitions. Such generality is needed in proving delicate embedding theorems for $V^v \hookrightarrow M^m$ in the dimensional range $2m > 3v$, in contrast to $2m > 4v$ for Whitney's theorem. The book is also set apart from the others we have considered by the attention given to degrees of differentiability. The reader may for example like to try and prove (a) every $\mathcal{C}^r$-manifold is $\mathcal{C}^r$-diffeomorphic to a $\mathcal{C}^\infty$-manifold ($1 \leqslant r < \infty$) and (b) if two $\mathcal{C}^s$-manifolds are $\mathcal{C}^r$-diffeomorphic they are also $\mathcal{C}^s$-diffeomorphic ($1 \leqslant r < s \leqslant \infty$). More delicate is the following: partitions of unity are of no use for constructing *analytic* approximations to a given map and more subtle techniques for passing from local to global are needed. If $\mathcal{C}^\omega$ denotes the class of analytic functions, then a deep theorem of Grauert (1958) states that, at least if V is compact, $\mathcal{C}^\omega(V, M)$ is dense in $\mathcal{C}^\infty(V, M)$. This is very fortunate since it implies that $\mathcal{C}^\omega$-theory agrees with $\mathcal{C}^\infty$-theory so far as *open* questions such as manifold classification, embeddings and immersions are concerned. Along with Bröcker and Jänich (1982), Hirsch is the author most focussed on differential topology, having no sections devoted to either Riemannian geometry or de Rham theory. However the reader anxious to fill out all the details of our sketch proof of the de Rham isomorphism theorem at the end of Chapter 7 could very usefully read Hirsch (1976), Chapter 6 on Morse theory.

LITERATURE REFERENCES

Th. Bröcker and K. Jänich, *Introduction to Differential Topology,* Cambridge University Press, 1982.

R. Bott and L. W. Tu, *Differential Forms in Algebraic Topology,* Springer, Heidelberg, 1982.

L. Conlon, *Differentiable Manifolds,* 2nd edition, Birkhäuser Advanced Texts, Basel, 2001.

R. Hamilton, The inverse function theorem of Nash and Moser, *Bull. Amer. Math. Soc.* **7**, 65–222, 1982.

M. W. Hirsch, *Differential Topology,* Springer, Heidelberg, 1976.

S. Lang, *Differential and Riemannian Manifolds,* Springer, Heidelberg, 1995.

I. Madsen and J. Tornehave, *From Calculus to Cohomology,* Cambridge University Press, 1997.

GENERAL REFERENCES

M. F. Atiyah, *The Moment Map in Symplectic Geometry,* Durham Symposium on Global Riemannian Geometry, Ellis Horwood Ltd., 1982.

M. A. Armstrong, *Basic Topology,* 5th corrected printing, Springer, Heidelberg, 1997.

D. Barden, Simply connected 5-manifolds, *Annals of Math.* **82**, 365–38, 1965.

Th. Bröcker and K. Jänich, *Introduction to Differential Topology,* Cambridge University Press, 1982.

Th. Bröcker and T. tom Dieck, *Representations of Compact Lie Groups,* Springer, Heidelberg, 1985.

W. Browder, *Surgery on Simply Connected Manifolds,* Ergebnisse der Math. und ihre Grenzgebiete **65**, Springer, Heidelberg, 1972.

J. Cerf, Topologie de certains espaces de plongement, *Bull. Soc. Math. France* **89**, 227-380, 1961.

C. Chevalley, *Theory of Lie Groups I,* Princeton University Press, Princeton NJ, 1946.

C. Chevalley and S. Eilenberg, Cohomology theory of Lie groups and Lie algebras, *Trans. Amer. Math. Soc.* **63**, 85-124, 1948.

J. Dieudonné, *Foundations of Modern Analysis,* Academic Press, New York, 1969.

Y. Eliashberg and N. Mishachev, *Introduction to the h-Principle,* AMS Graduate Studies in Math. **48**, 2002.

S. Eilenberg and N. Steenrod, *Foundations of Algebraic Topology,* Princeton University Press, 1952.

H. Grauert, On Levi's problem and the imbedding of real analytic manifolds, *Annals of Math.* **68**, 460–472, 1958.

M. Gromov, *Partial Differential Relations,* Springer, Heidelberg, 1986.

R. Hamilton, The inverse function theorem of Nash and Moser, *Bull. Amer. Math. Soc.* **7**, 65–222, 1982.

S. Helgason, *Differential Geometry, Lie Groups and Symmetric Spaces,* Academic Press, New York, 1978.

M. W. Hirsch, Immersions of manifolds, *Trans. Amer. Math. Soc.* **93**, 242–276, 1959.

M. W. Hirsch, *Differential Topology,* Springer, Heidelberg, 1976.

F. Hirzebruch and K-H. Mayer, *$O(n)$-Mannigfaltigkeiten, Exotische Sphären und Singularitäten,* Springer Lecture Notes **57**, Heidelberg, 1968.

D. Husemoller, *Fibre Bundles,* 3rd edition, Springer, Heidelberg, 1991.

K. Jänich, *Linear Algebra,* Springer, Heidelberg, 1994.

J. L. Kelley, *General Topology,* Springer, Heidelberg, 1977.

S. Lang, *Differential and Riemannian Manifolds,* Springer, Heidelberg, 1995.

J. Milnor *Differential Topology,* Princeton University Lecture Notes, 1958.

J. Milnor, *Morse Theory,* Princeton University Press, Princeton NJ, 1963.

J. Milnor *Lectures on the h-Cobordism Theorem,* Math. Notes **1**, Princeton Unversity Press, Princeton NJ, 1965.

J. Milnor, *Topology from a Differentiable Viewpoint,* University of Virginia Press, VA, 1965.

J. Nash The imbedding problem for Riemannian manifolds *Annals of Math.* **63**, 20–63, 1956.

P. Orlik and P. Wagreich, Seifert n-Manifolds, *Invent. Math.* **28**, 137–159, 1975.

G. de Rham, *Differentiable Manifolds,* Translated from the French by F.R. Smith, Springer, Heidelberg, 1984.

H. L. Royden, *Real Analysis, 2nd Edition,* Macmillan, New York, 1968.

H. Samelson, On de Rham's theorem, *Topology* **6**, 427–432, 1967.

G. P. Scott, The geometries of 3-manifolds, *Bull. London Math. Soc.* **15**, 401–487, 1983.

S. Smale, Generalized Poincaré's conjecture in dimensions greater than four, *Annals of Math.* **74**, 391-406, 1961.

N. E. Steenrod, *The Topology of Fibre Bundles,* Princeton University Press, Princeton NJ, 1951.

R. Stong, *Notes on Cobordism Theory,* Math. Notes **7**, Princeton University Press, Princeton NJ, 1968.

W. Thurston, Three-dimensional manifolds, Kleinian groups and hyperbolic geometry, *Bull. Amer. Math. Soc.* **6**, 357–381, 1982.

C. T. C. Wall, All 3-manifolds imbed in 5-space, *Bull. Amer. Math. Soc.* **71**, 564-567, 1965.

C. T. C. Wall, *Surgery on Compact Manifolds,* 2nd edition, AMS Surveys and Monograph, AMS, 1999.

F. W. Warner *Foundations of Differential Manifolds and Lie Groups,* Springer, Heidelberg, 1983.

INDEX